United States Department of Agriculture

I0511367

Economic Research Service
www.ers.usda.gov

Access this report online:

www.ers.usda.gov/publications/err-economic-research-report/err184

Download the charts contained in this report:

- Go to the report's index page www.ers.usda.gov/publications/
  err-economic-research-report/err184
- Click on the bulleted item "Download err184.zip"
- Open the chart you want, then save it to your computer

Recommended citation format for this publication:

Livingston, Michael, Jorge Fernandez-Cornejo, Jesse Unger, Craig Osteen, David Schimmelpfennig, Tim Park, and Dayton Lambert. *The Economics of Glyphosate Resistance Management in Corn and Soybean Production,* ERR-184, U.S. Department of Agriculture, Economic Research Service, April 2015.

United States Department of Agriculture

Economic
Research
Service

Economic
Research
Report
Number 184

April 2015

The Economics of Glyphosate Resistance Management in Corn and Soybean Production

Michael Livingston, Jorge Fernandez-Cornejo, Jesse Unger, Craig Osteen, David Schimmelpfennig, Tim Park, and Dayton Lambert

Abstract

Glyphosate, known by many trade names, including Roundup, is a highly effective herbicide. Widespread glyphosate use for corn and soybean has led to glyphosate resistance, which is now documented in 14 weed species affecting U.S. cropland, and recent surveys suggest that acreage with glyphosate-resistant (GR) weeds is expanding. Data from USDA's Agricultural Resource Management Survey (ARMS), along with the Benchmark Study (conducted independently by plant scientists), are used to address several issues raised by the spread of GR weeds. Choices made by growers that could help manage glyphosate resistance include using glyphosate during fewer years, combining it with one or more alternative herbicides, and, most importantly, not applying glyphosate during consecutive growing seasons. As a result, managing glyphosate resistance is more cost effective than ignoring it, and after about 2 years, the cumulative impact of the returns received is higher when managing instead of ignoring resistance.

Keywords: glyphosate, Roundup, corn, soybean, common property, resistance management practices, weeds, horseweed

Acknowledgments

The authors thank the following people for their comments: Marca Weinberg, James MacDonald, and Pat Sullivan of USDA's Economic Research Service; Linda Abbott of USDA's Office of the Chief Economist, Office of Risk Assessment and Cost-Benefit Analysis; Rosalind James of USDA's Agricultural Research Service, Crop Production and Protection; Michael Schechtman, Harold Coble (retired), and Jill Schroeder of USDA's Agricultural Research Service, Office of Pest Management Policy. In addition, George Frisvold of the University of Arizona; Terrance Hurley of the University of Minnesota; and Paul Mitchell of the University of Wisconsin provided peer reviews. We also thank Maria Williams for editorial assistance and Cynthia A. Ray for graphics and layout.

About the Authors

Michael Livingston is a principal with Promontory Financial Group LLC. This research was conducted while he was an economist with the USDA, Economic Research Service (ERS). The ideas, methodologies, views, and opinions expressed in this report are not those of Promontory Financial Group, LLC. Jorge Fernandez-Cornejo, Craig Osteen, David Schimmelpfennig, and Tim Park are USDA, ERS agricultural economists. Jesse Unger (currently at UC Berkeley) was an intern at USDA, ERS. Dayton Lambert is an associate professor at the University of Tennessee.

ii

The Economics of Glyphosate Resistance Management in Corn and Soybean Production, ERR-184
Economic Research Service/USDA

Contents

iii

The Economics of Glyphosate Resistance Management in Corn and Soybean Production, ERR-184
Economic Research Service/USDA

United States Department of Agriculture

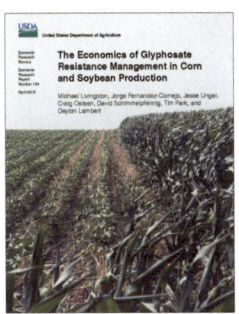

Find the full report at *www.ers.usda. gov/publications/err-economic-research-report/err184*

The Economics of Glyphosate Resistance Management in Corn and Soybean Production

Michael Livingston, Jorge Fernandez-Cornejo, Jesse Unger, Craig Osteen, David Schimmelpfennig, Tim Park, and Dayton Lambert

What Is the Issue?

Glyphosate—usually known by many trade names, including Roundup—has been the most widely used herbicide in the United States since 2001. It effectively controls many weed species, and generally costs less than the herbicides that it replaced. Because several major crop varieties have been genetically engineered to tolerate glyphosate, crop growers can spray entire fields planted to glyphosate-tolerant (GT) varieties, killing the weeds but not the crops. This practice makes it easier to manage weeds using less tillage, which can help reduce soil erosion and improve soil quality and water conservation. However, glyphosate's effectiveness is declining as weed resistance mounts—14 glyphosate-resistant (GR) weed species currently affect U.S. crop-production areas. GR weeds can reduce crop yields and increase weed-control costs, and recent surveys suggest that the amount of affected cropland is increasing. This study addresses several issues raised by the spread of GR weeds and the effect on U.S. agriculture.

What Did the Study Find?

Reliance on glyphosate, by many growers, as the sole herbicide to control weeds is the primary factor underlying the evolution of GR weeds. Using glyphosate in isolation can select for glyphosate resistance by controlling susceptible weeds while allowing more resistant weeds to survive, which can then propagate and spread. Using herbicides with different modes of action, which affect susceptible weeds differently, and rotating their use over time can result in fewer herbicide-resistant weeds.

Growers report that glyphosate resistance is more prevalent in soybean production than in corn production. Since the commercial introduction of GT crops in 1996, glyphosate use in soybean production has promoted the spread of GR weeds more than its use in corn production. In surveys of crop production practices, growers were asked to report their concerns about glyphosate resistance, either as the presence of "GR weeds" in corn or "declines in glyphosate effectiveness" in soybeans. They reported GR-weed infestations on 5.6 percent of the corn acres in 2010 and declines in glyphosate effectiveness in about 40 percent of soybean acres in 2012, with the majority of those acres in the Corn Belt and Northern Plains.

Soybean production relies more on glyphosate than does corn production. While more herbicide active ingredient was applied to corn than to soybeans, herbicides other than glyphosate accounted for most of the herbicide applied to corn acres. In addition, tillage, which controls weeds without promoting herbicide resistance, was used on a greater share of corn than soybean acreage, whereas no-till was used on a greater share of soybeans than corn acreage. Other findings show:

- GT varieties were planted to more soybean than corn acres;

ERS is a primary source of economic research and analysis from the U.S. Department of Agriculture, providing timely information on economic and policy issues related to agriculture, food, the environment, and rural America.

www.ers.usda.gov

- Much more glyphosate (pounds of active ingredient) was applied to soybean than to corn fields;
- Glyphosate was used on more soybean than corn acres;
- Herbicide use practices were consistent with glyphosate-resistance management on fewer soybean acres (60 percent) than on corn acres (82 percent); and
- Glyphosate-resistance management was more likely to be done proactively on corn acres and more likely to be done reactively, in response to GR weeds, on soybean acres.

Managing glyphosate resistance is more cost effective than ignoring resistance. Simulation results over a 20-year period show that herbicide choices that help manage glyphosate resistance differ from short-term herbicide choices that ignore glyphosate resistance in three important ways. Choices that manage resistance (1) use glyphosate during fewer years; (2) often combine glyphosate with one or more alternative herbicides; and (3) most importantly, avoid applying glyphosate in consecutive growing seasons. As a result, glyphosate resistance is managed more cost effectively, and after about 2 consecutive years of managing resistance, the cumulative impact of the returns received   exceeds that received when ignoring resistance.

Corn and soybean growers responded similarly to the reported presence of GR weeds or declines in glyphosate effectiveness. The most common survey response—consistent with glyphosate-resistance management—was to use other herbicides in addition to glyphosate. Growers used this practice on over 84 percent of corn acres with GR weeds and on 71 percent of soybean acres with reduced glyphosate effectiveness. The next most common response was to increase the amount of glyphosate used. Growers used this practice on 25 percent of corn acres with GR weeds and 39 percent of soybean acres with reduced glyphosate effectiveness.

Corn growers who reported GR weeds and soybean growers who reported reduced glyphosate effectiveness realized lower returns than similar corn and soybean growers who did not report them. In addition, corn and soybean growers who used glyphosate alone received lower yields and returns than similar corn and soybean growers who used at least one other herbicide in combination with glyphosate. Although the crop growers using more than one herbicide had higher production costs, the additional costs were more than offset by higher yields.

Economic incentives encourage cooperative use of resistance management practices (RMPs). Simulations show that weed-seed dispersal from a field where crop growers ignore resistance when managing weeds could reduce the returns on nearby fields, and that the reduction could be larger for a nearby field where growers manage resistance than where they ignore it. This result suggests that corn and soybean growers have an economic incentive to encourage neighbors to use RMPs and may also be aware of the incentive. Some soybean growers in Arkansas and North Carolina are responding to such incentives and collaborating in the management of herbicide-resistant weeds.

How Was the Study Conducted?

This research relies on two primary data sources on corn and soybean production: (1) the Agricultural Resource Management Survey (ARMS), managed jointly by USDA's Economic Research Service and National Agricultural Statistics Service, and (2) the Benchmark Study conducted by university plant scientists and sponsored by Monsanto Company. Corn and soybean production is the focus because these are the leading U.S. crops in terms of planted acres, and growers of these crops are currently managing GR weeds.

To examine economic factors shaping the herbicide-use decisions of crop growers, we combined a biological model of weed growth and glyphosate resistance with an economic model of weed management to create a bio-economic model to compare herbicide choices that (1) help manage glyphosate resistance and maximize longrun returns and (2) ignore glyphosate resistance and instead maximize shortrun returns.

The Economics of Glyphosate Resistance Management in Corn and Soybean Production

Introduction

Glyphosate—known by many trade names, including Roundup—is an herbicide effective for controlling many weed species. Because alfalfa, canola, corn, cotton, soybean, and sugar beet varieties have been genetically engineered to tolerate glyphosate, crop growers can spray entire fields planted to glyphosate-tolerant (GT) varieties, killing weeds but not the crops. This practice makes it easier to manage weeds using less tillage (Fawcett, 2002; NRC, 2010), which can help reduce soil erosion and improve soil quality and water conservation (Reeves, 1994, 1997; Kaspar et al., 2001; NRC, 2010; Price et al., 2011; Fernandez-Cornejo et al., 2014a).

Glyphosate was first marketed in 1974 under the name Roundup. Its use increased rapidly with the commercial introduction of GT corn, soybeans, and cotton in 1996 and patent expiration in 2000, which led to the availability of relatively inexpensive generic equivalents. Glyphosate is reported to be less toxic and less persistent in the environment relative to the herbicides that it replaced (Malik et al., 1989, Duke and Powles, 2008; NRC, 2010). The National Research Council (2010) reported that glyphosate is biodegraded by soil bacteria and it has a very low toxicity to mammals, birds, and fish.[1] As a result, glyphosate has been the most widely used pesticide in the United States since 2001 (Fernandez-Cornejo et al., 2014; Grube et al., 2011; Osteen and Fernandez-Cornejo, 2013). U.S. crop growers now plant 93 percent of their soybean acres and 85 percent of their corn acres to genetically engineered (GE) herbicide-tolerant (HT) varieties (fig. 1).[2] The emergence of the HT varieties led corn and soybean growers to increase their use of glyphosate over time and reduce their use of all other herbicides. During 1996-2003, herbicide use in corn and soybean production declined from about 293 million pounds of active ingredient to around 247 million pounds. Since 2003, herbicide use on acreage planted with these two crops has increased to almost 353 million pounds of active ingredient in 2013, with glyphosate accounting for over 57 percent of the total.[3]

---

[1]Regarding toxicity to humans and based on risk assessments, the U.S. Environmental Protection Agency (EPA) concluded that glyphosate had minimal human dietary exposure and risk, and that exposure to workers and other applicators would not pose undue risk, because of low acute toxicity. However, EPA recommended personal protective equipment for skin and eye irritation, and a 12-hour re-entry interval on treated agricultural areas to mitigate potential risks (U.S. Environmental Protection Agency, 1993). Glyphosate is currently under a standard registration review by EPA, the outcome of which is expected in 2015. (http://www.epa.gov/oppsrrd1/registration_review/reg_review_status.htm). More recently, on March 20, 2015, the International Agency for Research on Cancer (IARC) of the World Health Organization (WHO) issued a summary of the final evaluations of 5 pesticides including glyphosate that classifies glyphosate as "Probably Carcinogenic to Humans" http://www.iarc.fr/en/media-centre/iarcnews/pdf/MonographVolume112.pdf

[2]Genetically engineered corn and soybean varieties that tolerate the herbicide glufosinate became commercially available after 1996, but the majority of herbicide-tolerant (HT) corn and soybean varieties planted in the United States have been glyphosate-tolerant (GT) varieties.

[3]The increase in herbicide use in corn and soybean production is partially due to the increase in acres planted to these two crops—from 152 million acres in 2003 to 175 million acres in 2013.

1

The Economics of Glyphosate Resistance Management in Corn and Soybean Production, ERR-184
Economic Research Service/USDA

Figure 1

Adoption of genetically engineered herbicide-tolerant (HT) corn and soybeans and pounds of herbicide active ingredient (a.i.) applied to those crops, 1996-2013

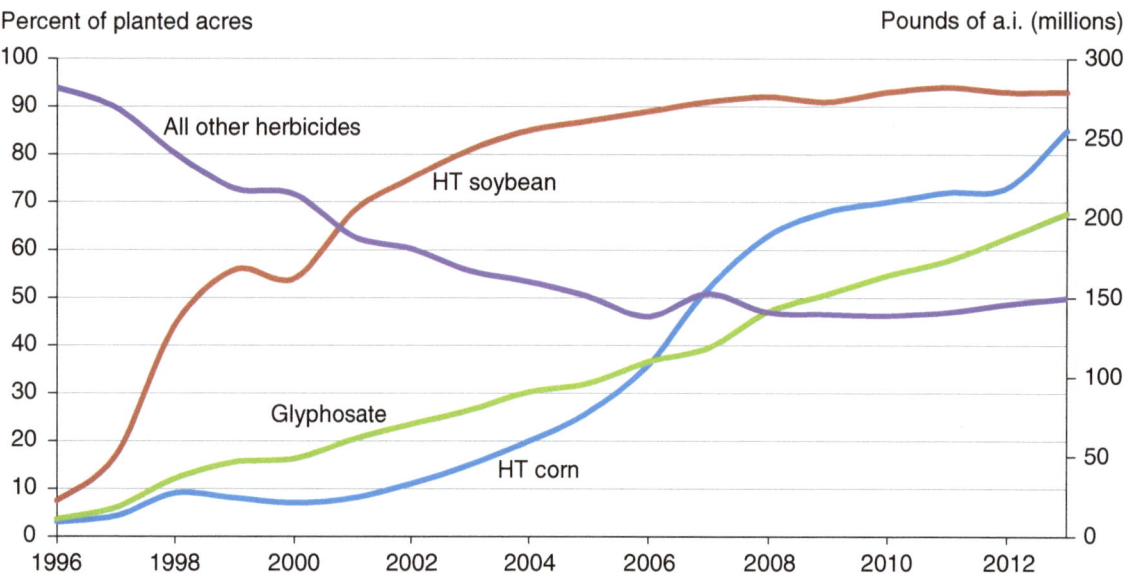

Source: USDA, Economic Research Service estimates. Data for GE adoption from Fernandez-Cornejo, J. Adoption of Genetically Engineered Crops in the U.S., Data Product, http://www.ers.usda.gov/Data/BiotechCrops/ July 2014.
Data for herbicide use estimates come from the Agricultural Resource Management Survey (ARMS) Phase II and from USDA (2014). Regressions were used to interpolate herbicide missing data.

However, glyphosate is becoming less effective at controlling some weeds. The International Survey of Herbicide Resistant Weeds identified 14 glyphosate-resistant (GR) weed species currently affecting U.S. crop-production areas (Heap, 2014).[4] GR weeds can increase weed control costs and decrease crop yields (Shaw et al., 2011; Mueller et al., 2005; Scott and VanGessel, 2006; Culpepper et al., 2008; Culpepper and Kichler, 2009; Webster and Sosnoskie, 2010). Recent surveys of crop growers in 31 States suggest that acreage with GR weeds is increasing (Fraser, 2013). Because no new major herbicide active ingredients have become commercially available in the last 20 years, and because few new herbicides are expected to be available anytime soon (Harker et al., 2012), plant scientists have suggested that slowing the spread of GR weeds is a serious challenge facing U.S. crop growers (NRC, 2010).

Exclusive reliance on one herbicide as the sole control tactic over the majority of the corn, cotton, and soybean acres in the Midwest and South is believed to be the main factor underlying the growing resistance of weeds to glyphosate. The Weed Science Society of America (WSSA) recently described an herbicide resistance management strategy that can reduce the spread of weeds resistant to a single herbicide (Norsworthy et al., 2012). The strategy involves understanding the biology of the weeds that are present and using diverse chemical, cultural, and mechanical methods to control weeds and reduce the production and dissemination of weed seeds. Using multiple herbicides with different modes of action (MOA), rotating use of different herbicide MOAs over time, and adopting several other resistance management practices (RMPs) can reduce the spread of weeds resistant to a single herbicide.[5]

---

[4]The current U.S. count is 14 according to the International Survey of Herbicide Resistant Weeds. http://weedscience.com/summary/MOA.aspx (accessed on January 5, 2015).

[5]Herbicides control weeds by disrupting one or more vital metabolic process, like photosynthesis or protein synthesis. Herbicides that kill weeds by disrupting different metabolic processes are said to have different modes (or mechanisms) of action, or MOA (see Glossary).

2

The Economics of Glyphosate Resistance Management in Corn and Soybean Production, ERR-184
Economic Research Service/USDA

Perhaps, because of a lack of information and economic incentives, many crop growers only use *some* RMPs, and they use them *after resistance develops*, in response to resistance, instead of using them to delay the onset and spread of herbicide resistance. Many crop growers reportedly believe that there is no need to use some RMPs, holding to a view that new herbicides will be available in time to control weeds resistant to currently available herbicides (Norsworthy et al., 2012). Some RMPs also increase current production costs, whereas the future benefits of delaying the spread of weed resistance by using RMPs are uncertain (Frisvold et al., 2009).

Moreover, because weed seeds can disperse between fields by wind (Dauer et al., 2009), water, animals, and humans, including movement on farm equipment (Ross and Lembi, 2009), the effectiveness of RMPs in delaying resistance on one farm can depend on their use on neighboring farms. As a result, the susceptibility of weeds to glyphosate is an example of what economists refer to as a "common pool resource" (see Glossary), which is especially prone to overuse (Ostrom et al., 1999).[6] For mobile pests like insects and weeds, market-based economic incentives might be insufficient to ensure that resistance is managed in an economically optimal manner (Miranowski and Carlson, 1986; Livingston, 2013).

Glyphosate-Resistance Management Strategies

Stakeholders are responding to the spread of GR weeds in a number of ways. Plant scientists focus on information campaigns that raise awareness of resistance and communicate the benefits of using RMPs (Duke and Powles, 2009; Norsworthy et al., 2012; Price et al., 2011; Vencill et al., 2012). The seed industry, government agencies, and other organizations are helping to finance these efforts (Farm Industry News, 2014; Boerboom and Owen, 2006). The seed industry is in the process of registering new GE crop varieties that can tolerate herbicides with a range of MOAs. GE corn varieties that tolerate glyphosate and glufosinate are currently available, and USDA has been reviewing industry proposals to deregulate new GE corn and soybean varieties that tolerate glyphosate and 2,4-D (now deregulated) or glyphosate and dicamba (Johnson et al., 2012; APHIS, 2014).[7] Some herbicide registrants have been offering incentives (such as per-acre payments) to purchase specific herbicide products with different MOAs than glyphosate's, such as acetochlor, atrazine, or other chemicals, to use with glyphosate on corn, cotton, or soybeans, depending on the type and brand of seed used.

USDA's Natural Resources Conservation Service (NRCS) is promoting the use of RMPs under its Integrated Pest Management Herbicide Resistance Weed Conservation Plan. The agency provides financial assistance for developing conservation activity plans under the Environmental Quality Incentives Program. The plans provide guidelines to delay herbicide resistance and meet soil, water, and air quality objectives.

---

[6]Some economists consider pest susceptibility to pesticides a common pool resource. When a pest moves from farm to farm, the pest control decisions made by any given farmer will affect the pest susceptibility (and returns) accruing to that farmer as well as those accruing to nearby farmers, but to a lesser extent. However, the effects of any one farmer's control decisions on the (susceptibility of) regional pest populations are practically negligible and the benefits and costs associated with those effects are not borne by any given farmer. Thus, those effects might not be accounted for in the farmer's control decision (Feder and Regev, 1975; Fernandez-Cornejo et al., 2014a). Pollen and seeds of many different weed species can disperse between farms in the air and in conjunction with the movement of animals and farm machinery (Fernandez-Cornejo et al., 2014a).

[7]Concerns have been raised that the longrun sustainability of these new GE varieties might be undermined by the lack of information and economic incentives that are currently keeping crop growers from using RMPs until resistance occurs (Union of Concerned Scientists, 2013).

3

The Economics of Glyphosate Resistance Management in Corn and Soybean Production, ERR-184
Economic Research Service/USDA

In this report, we use new data from (1) the Agricultural Resource Management Survey (ARMS) and (2) the Benchmark Study (Shaw et al., 2011) to address several questions raised by the spread of GR weeds (see box, "Data Sources"):

1. How have corn and soybean growers managed weeds since GT crops became commercially available? Have weeds been managed differently for these two crops? What are the implications for managing glyphosate resistance?

2. How do herbicide choices that help manage glyphosate resistance and maximize longrun returns differ from herbicide choices that ignore resistance and instead maximize shortrun returns? Will an understanding of the differences in these choices help identify potential barriers to using RMPs and promoting their use?

3. How do the weed-management practices currently used in corn and soybean production compare to weed-management choices that maximize longrun returns?

4. What are the current effects of glyphosate resistance on production costs, crop yields, and returns? What are the current effects of using glyphosate by itself, which can exacerbate glyphosate resistance, relative to using glyphosate with at least one different herbicide MOA? Do the shortrun economic benefits of using glyphosate by itself outweigh the longrun costs associated with the spread of GR weeds?

5. How does the common-pool nature of weed susceptibility to glyphosate affect RMP use and production costs, yields, and returns? What are the implications for the use of RMPs in corn and soybeans?

Data Sources

ARMS data. This research relies on the U.S. Department of Agriculture (USDA) Agricultural Resource Management Survey (ARMS), which the USDA, Economic Research Service (ERS) and National Agricultural Statistics Service (NASS) jointly administer. Enumerator-assisted surveys of farmers are conducted in three phases. Phase I is a screening questionnaire used to verify that respondents meet certain criteria. Phase II, conducted during the fall of the reference year, focuses on operations that produce specific crops. Currently, up to two crops are surveyed each year. A specific field planted to the crop is chosen at random for questions about yield, input use and costs, and production practices.

Phase III, conducted in the winter following the reference year, requests information on whole-farm revenues and expenses and characteristics of the farm operators and their households and businesses. Commodity versions of Phase III surveys request information on prices received and total expenditures for the crops surveyed in Phase II, which is used to estimate production costs and returns. Many, but not all, of the crop growers who respond to the Phase II survey also respond to the commodity version of the Phase III survey.

We focused on corn and soybean production in the United States in this report because they are the leading U.S. crops in terms of planted acres. Furthermore, the growers of these two crops are currently managing glyphosate-resistant (GR) weeds, and we had new data for these two crops. We used Phase II data collected from corn (1996-2001, 2005, and 2010) and soybean (1996-2000, 2002, 2006, and 2012) growers to examine trends in the use of genetically engineered (GE) herbicide-tolerant (HT) varieties, herbicide use, and other weed-management practices.

The Phase II corn (2010) and soybean (2012) questionnaires also collected information about the prevalence of and responses to GR weeds. We merged these data with data from the Phase III corn and soybean commodity versions of ARMS to estimate the impacts of GR weeds on corn and soybean production costs, yields, and returns.

NASS uses a stratified sampling strategy to improve the reliability of estimates based on ARMS. Sample-selection probabilities vary by farm size, geographic area, and commodities produced. For example, larger operations are more likely to be sampled than smaller operations. Population estimates are produced by weighting sample observations to account for their probability of selection in the sample, unless incomplete subsets of the data are examined, as is the case when we compare specific groups of surveyed crop growers.

Benchmark Study data. Plant scientists recently examined the effects of resistance management practices (RMPs) on crop yield, weed density, and returns to weed management (Shaw et al., 2011). Corn, cotton, and soybean growers in six States (IL, IN, IA, MS, NC, and NE) took part in the study, and each grower was asked to pick a field on his or her farm and split the field into two halves. On the first half, each grower was asked to manage weeds as usual and, on the second half, to use an RMP. On both field halves, data were collected on yield, weed density, herbicide use, tillage practices, irrigation use or nonuse, the field's latitude, and the type of seed that was planted during 2006-2010.

We used the data collected from study participants who rotated GT corn and GT soybean and planted continuous GT corn and planted continuous GT soybeans during 2006-2009—excluding 2010 because of data quality issues and participant attrition. We used the data to estimate and then specify empirical relationships between crop yield and annual weed density in simulation models to examine the characteristics of different herbicide-use strategies.

5

The Economics of Glyphosate Resistance Management in Corn and Soybean Production, ERR-184
Economic Research Service/USDA

Weed-Management Practices in Corn and Soybean Production, 1996-2012

To identify ways to promote the use of RMPs in corn and soybean production, it is important to understand how growers have managed weeds in those crops since the commercial introduction of GT varieties in 1996. We examined ARMS data collected from corn and soybean growers during 1996-2012 to identify relevant trends and differences in corn and soybean management practices and explanations for the trends and differences.

Since using glyphosate by itself repeatedly over time is the most important factor underlying the evolution of glyphosate resistance (Norsworthy et al., 2012), ARMS data suggest that herbicide-use practices in soybean production promoted the spread of GR weeds to a greater extent than herbicide-use practices in corn production. HT varieties (mostly GT varieties) were planted on more soybean than corn acres (see fig. 1); much more glyphosate (expressed in pounds of active ingredient) was applied to soybean than to corn fields (fig. 2); and glyphosate was used by itself on far more soybean than corn acres (fig. 3). At the same time, the total quantity of herbicide active ingredient applied was much greater on corn than soybean acreage, and herbicides other than glyphosate accounted for the majority of the herbicides applied to corn. In addition, tillage, which controls weeds without encouraging herbicide resistance, was used more in corn than in soybean production, whereas no-till production systems were used more for soybeans than for corn (Horowitz et al., 2010).

Weed management in corn fields involves not only glyphosate, but also other inexpensive herbicides, such as atrazine. In contrast, weed management in soybean fields is largely managed with glyphosate alone, because the next best alternative herbicides to control soybean weeds, especially broadleaf weeds, are more expensive, less effective, and can injure soybean plants (NRC, 2010). These facts help explain why GT soybean adoption was more rapid than GT corn adoption, why much more glyphosate was used in soybean than in corn production, and why far more soybean than corn acres received glyphosate by itself during 1996-2012.

Herbicide Use

Herbicide use on soybeans in surveyed States increased from about 60 million pounds of active ingredient (a.i.) in 1996 to 103 million pounds in 2006 (see fig. 2). Glyphosate's share increased from 15 percent of the herbicide active ingredient applied in 1996 to 55 percent in 2000; by 2006, its share had increased to 89 percent. The percentage of soybean acres (in surveyed States) treated with glyphosate, by itself or in combination with other herbicides, increased from about 25 percent in 1996 to over 60 percent in 2000, and to about 95 percent in 2006 (fig. 3). Moreover, soybean acres treated with glyphosate as the sole herbicide increased from only 9 percent in 1996 to 73 percent in 2006.

Because of the presence of GR weeds in soybean fields, discussed later, both of these trends changed between 2006 and 2012. The amount of other herbicides (with different MOAs) applied to soybeans almost doubled, from 11.4 million pounds in 2006 to 22.5 million pounds in 2012 (see fig. 2).[8] As a result, glyphosate accounted for 82 percent of total herbicide active ingredient applied to soybeans in 2012, down from 89 percent. The soybean acreage that received glyphosate by itself also declined,

---

[8]Weed scientists classify glyphosate as the only inhibitor of 5-enolpyruvyl-shikimate-3-phosphate (EPSP) synthase, a unique mode of action (Mallory-Smith and Retzinger, 2003). All other corn and soybean herbicides considered in this report have a different mode of action than glyphosate.

6

The Economics of Glyphosate Resistance Management in Corn and Soybean Production, ERR-184
Economic Research Service/USDA

from 51 million in 2006 to 30 million in 2012 (a decline from 73 percent to 44 percent of glypho-sate-treated acreage), because the number of soybean acres that received glyphosate and at least one different herbicide MOA more than doubled, from 19 million acres in 2006 to over 38 million acres in 2012 (an increase from 27 percent to 56 percent of glyphosate-treated acreage) (see fig. 3).

Figure 2

Herbicide quantity applied to corn and soybean, surveyed States, 1996-2012

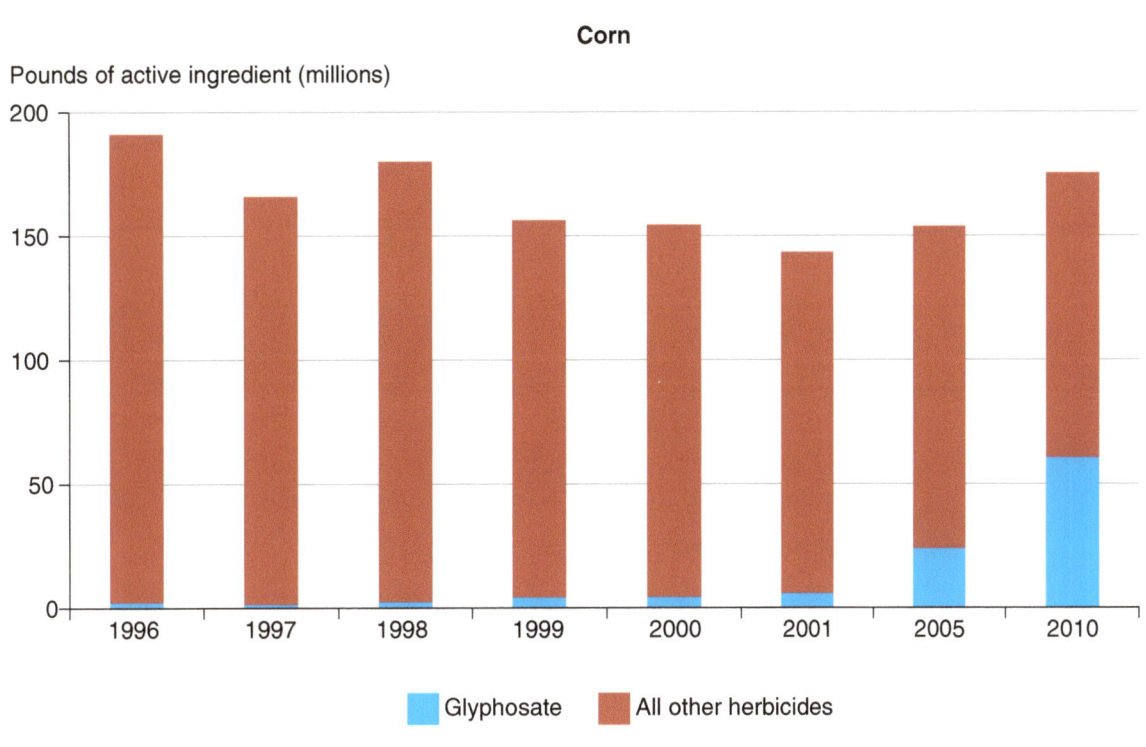

Corn

Pounds of active ingredient (millions)

Glyphosate   All other herbicides

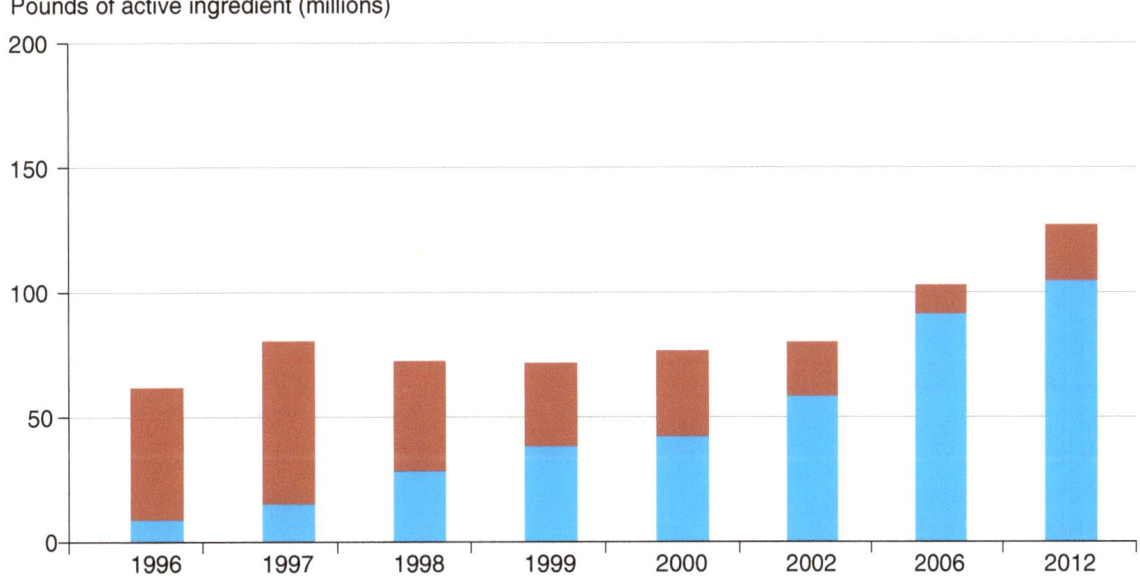

Soybeans

Pounds of active ingredient (millions)

Source: USDA, Economic Research Service estimates using data from the Agricultural Resource Management Survey (ARMS) Phase II.

The Economics of Glyphosate Resistance Management in Corn and Soybean Production, ERR-184
Economic Research Service/USDA

Figure 3

Corn and soybean acreage that received different groups of herbicides, surveyed States, 1996-2012

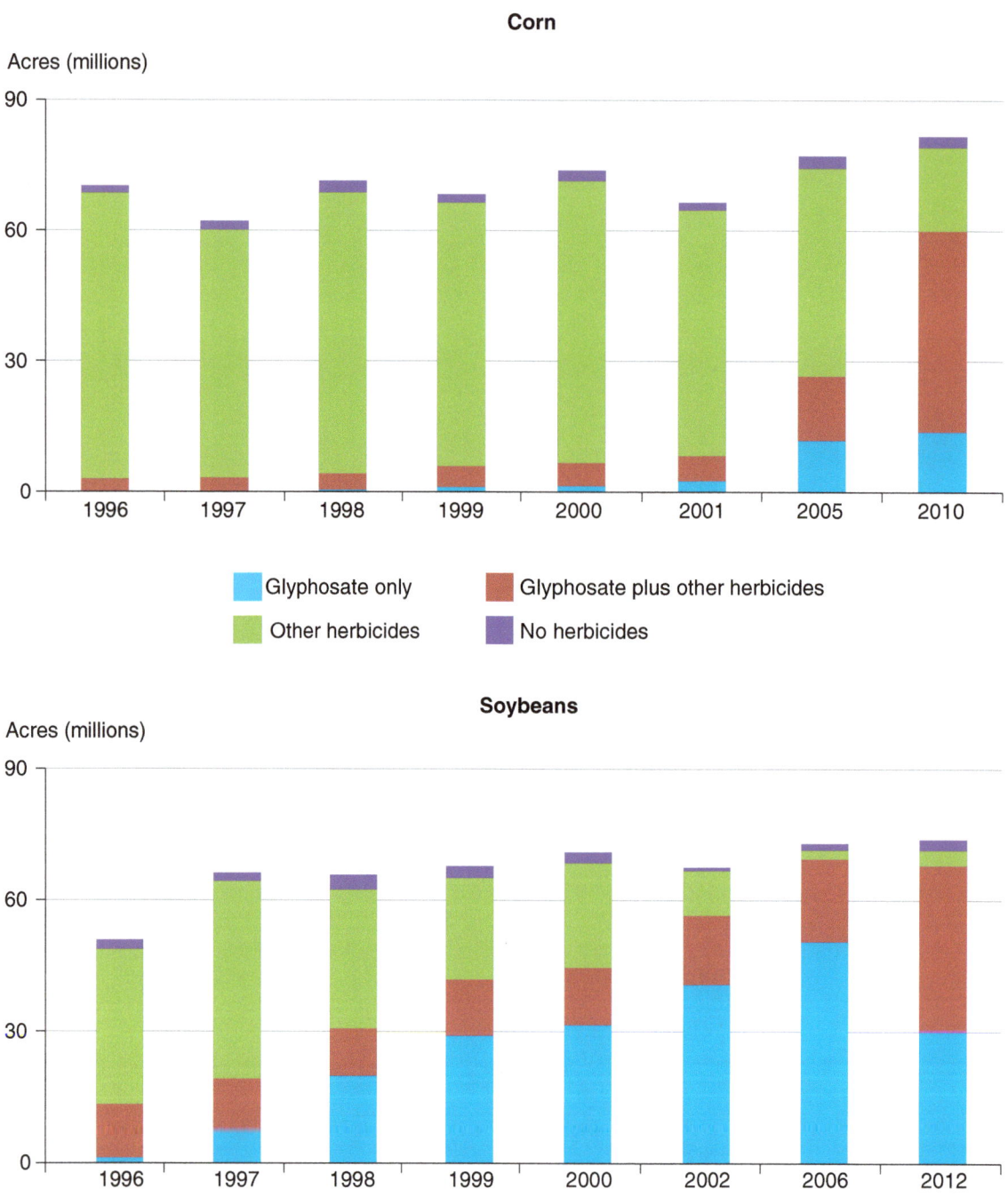

Source: USDA, Economic Research Service using data from the Agricultural Resource Management Survey (ARMS) Phase II.

Much more total herbicide was applied in corn than in soybean production, and herbicides other than glyphosate accounted for the majority of herbicides used on corn fields. Herbicide use on corn in surveyed States declined from 191 million pounds of active ingredient in 1996 to 143 million pounds in 2001, but then increased to 175 million pounds in 2010 (see fig. 2). Glyphosate accounted for only 1 percent of herbicide use in 1996, but as HT corn varieties were planted to more acres, glyphosate use grew to 35 percent of total herbicides applied in 2010.

The Economics of Glyphosate Resistance Management in Corn and Soybean Production, ERR-184
Economic Research Service/USDA

The percentage of glyphosate-treated corn acreage rose steadily from 4 percent of planted acres in 1996 to 35 percent in 2005 to 73 percent in 2010. So, the majority of surveyed corn acreage received no glyphosate from 1996 (96 percent) to 2005 (65 percent). The percentage of all glyphosate-treated corn acres that received only glyphosate increased from about 1 percent in 1996 to 21 percent in 2000 to 44 percent in 2005, but declined to only 23 percent in 2010, perhaps due to the presence of GR weeds (see fig. 3).[9] So, the majority of glyphosate-treated corn acres received at least one additional herbicide MOA throughout this period, with the lowest percentage occurring in 2005.

Resistance Management Practices

Corn and soybean growers used a variety of RMPs during 1996-2012. (Table 1 shows the use of a selection of those practices on corn and soybean acres based on ARMS data.) The majority of corn and soybean acres were scouted for weeds, exceeding 80 percent in most survey years, and this practice was used on an increasing percentage of soybean and corn acres, although corn acres scouted did not increase as much at the end of the period. Crops were rotated on over 75 percent of soybean acres and a smaller percentage of corn acres, with little change in either crop during this period.

Tillage was used on over half of corn and soybean acres, with a greater percentage of corn than soybean acres being tilled.[10] Except for soybeans in 2012, there was a slight downward trend in tilled acres for both crops, perhaps in response to GR weed infestations. As Price et al. (2011) observe, many acres under conservation tillage "are at risk of being converted to higher-intensity tillage systems due to the inability to control" glyphosate-resistant weeds. They add that "the decline of conservation tillage is inevitable without the development and rapid adoption of integrated, effective weed control strategies." Moreover, if conservation tillage declines, its benefits (in reducing soil erosion and improving soil quality and water conservation) will also be at risk.

The percentage of planted acres that received herbicides with MOAs other than glyphosate's declined at least initially, for both crops; however, the decline was much more dramatic for soybeans than for corn. The share of soybean acres receiving herbicides with MOAs other than glyphosate's declined steadily from 93 percent in 1996 to 29 percent in 2006, before increasing to 56 percent in 2012, perhaps due to the rising presence of GR weeds. The share of corn acres that received at least one herbicide MOA other than glyphosate's declined from 98 percent in 1996 to 80 percent in 2010.

Other RMPs were used on less than 50 percent of corn and soybean acres during this period. A greater percentage of corn than soybean acres was cultivated for weed control, but acres receiving that practice declined steadily for both crops. For both crops, the practice of rotating or alternating pesticides to delay pesticide resistance increased during the first half of the period, then declined during the latter half, with the exception of the last years, perhaps in response to the presence of more GR weeds. Mowing field edges and roadways to prevent pest introductions increased after 2000, as did two other RMPs, previously used sparingly: adjustments to planting dates (to avoid weeds) and plant density or row spacing (to crowd out weeds). Finally, some practices (not included in table 1) showing very small differences in their use across time and between corn and soybean acres were "planting a cover crop in the fall," on 0.9 percent of corn and 0.5 percent of soybean

---

[9]Over 1.4 million corn acres in surveyed States received glyphosate by itself in 2000, increasing to 13.8 million acres in 2010. Over 5.3 million corn acres received glyphosate plus at least one different herbicide MOA in 2000; this increased to 46.2 million acres by 2010.

[10]In the ARMS Phase II questionnaire, farmers were asked to indicate if they "plow down crop residue (using conventional tillage) with the purpose of reducing the spread of pests in the field." For this reason, answers to these questions may be somewhat different from usual tillage questions based on actual tillage operations.

9

The Economics of Glyphosate Resistance Management in Corn and Soybean Production, ERR-184
Economic Research Service/USDA

acres; "keeping written records of weeds observed," on 21 percent of corn and 21 percent of soybean acres; and "cleaning equipment between fields," on 29 percent of corn and 31 percent of soybean acres.

Table 1

Use of resistance management practices on corn and soybeans, 1996-2012

Year	Used herbicide other than glyphosate	Used tillage	Scouted for weeds	Rotated crops	Adjusted planting dates	Adjusted plant density	Rotated pesticides	Mowed field edges	Cultivated for weed control
				Planted acres (percent)					
				Corn					
1996	98	82	81	71	7	5	32	34	55
1997	97	83	80	75	4	5	32	34	55
1998	95	87	86	74	3	5	46	34	42
1999	96	85	86	76	4	5	43	40	48
2000	95	83	83	74	7	7	45	40	38
2001	94	84	84	77	3	4	40	31	37
2005	81	77	89	73	10	11	24	43	15
2010	80	74	88	69	14	13	27	44	15
				Soybean					
1996	93	66	79	81	8	11	28	31	29
1997	86	71	83	80	3	11	32	31	28
1998	65	71	85	84	5	11	43	31	26
1999	53	69	87	79	3	17	36	38	22
2000	52	69	85	77	5	17	34	41	17
2002	38	66	85	84	4	18	22	37	19
2006	29	56	90	85	13	19	13	45	8
2012	56	59	94	82	15	18	24	43	8

Note: In the ARMS Phase II questionnaire, growers were asked to indicate if they used each resistance management practice "with the purpose of reducing the spread of pests in the field."

Source: USDA, Economic Research Service using data from the Agricultural Resource Management Survey (ARMS) Phase II.

Managing Weeds: Short- and Long-Term Views

Soybean and corn acres that received glyphosate as the only herbicide treatment increased as more acres were planted to GT varieties, although this trend was much more pronounced in soybean than in corn production. Because corn production, unlike soybean production, benefited from inexpensive and effective alternative herbicides, the fact that the trend existed at all in corn production suggests that a lack of alternative herbicides was not the only underlying factor.

A potentially important factor encouraging use of glyphosate by itself is that adding other herbicides with different MOAs can increase weed management costs, while the benefits in terms of resistance management are uncertain (not precisely known by the grower). Using multiple herbicides with different MOAs and, more importantly, rotating them over time can be very effective to delay the spread of weed resistance. However, because weed seeds can disperse between fields by wind, water, animals, and human activity, including the movement of farm equipment, the effectiveness of these practices on one farm can depend on their use on nearby farms. As a result, weed susceptibility to glyphosate may decline in the absence of coordinated action by all neighboring farms (Ostrom et al., 1999).

Before exploring the potential effectiveness of policies promoting herbicide-use practices that delay glyphosate resistance, it is necessary to understand how those practices affect production costs, crop yields, and returns, and how those quantities might be affected by the potential for weed-seed dispersal between farms. To accomplish these objectives, we combined a biological model of weed growth and glyphosate resistance with an economic model of weed management.[11] In this bio-economic model, a representative crop grower is assumed to observe the weed-seed density and glyphosate-resistance level in a crop field at the beginning of each year.[12] The crop grower then selects one of the following herbicide choices: (1) a pre-emergence (residual) herbicide, plus post-emergence glyphosate; (2) residual herbicide and glyphosate, plus an alternative (also post-emergence) herbicide; (3) residual herbicide, plus an alternative herbicide; (4) just glyphosate; (5) glyphosate plus the alternative herbicide; or (6) just the alternative herbicide.[13]

The model simulates the effects of the herbicide choice on weed growth, which determine crop yield and returns received each year, and the following year's initial weed-seed density and resistance level.[14] We used the model to examine two choices: managing resistance and ignoring resistance. Managing resistance involves a long-term view accounting for the cost and yield effects of glyphosate resistance over time and maximizing the present value of annual returns received. Ignoring

---

[11]See Appendix 1 – Bio-Economic Optimization Model for a complete description of the model.

[12]We also examined a bio-economic model that simulates resistance to glyphosate and the alternative, post-emergence herbicide. Because the results are qualitatively similar to the results for the model that only simulates glyphosate resistance, we present only the results for the latter model for simplicity.

[13]Application rates and prices are based on 2010 ARMS data. Glyphosate is applied after the crop emerges (post-emergence) at 0.94 pounds per acre and $6.49 per pound. For corn, the alternative, post-emergence herbicide is 0.98 pounds of atrazine at $6.70 per pound. For soybeans, the alternative, post-emergence herbicide is 1.18 pounds of acetochlor, at $10.05 per pound. For both crops, the pre-emergence, residual herbicide is 1.41 pounds of acetochlor applied to the soil.

[14]In this report, *total returns* (or just *returns* for short) are per-acre revenues (gross value of production) minus per-acre operating costs (e.g., per-acre cost of fertilizer, pesticides, seed, energy, repairs, irrigation water) and per-acre overhead costs (e.g., opportunity costs of unpaid labor and land, taxes, insurance, and general farm overhead). Revenues per acre are equal to the crop yield times the crop price. We also use *operating returns*, which are equal to per-acre revenues minus per-acre operating costs.

11

The Economics of Glyphosate Resistance Management in Corn and Soybean Production, ERR-184
Economic Research Service/USDA

resistance is a short-term perspective that ignores glyphosate resistance and instead maximizes the current-year annual returns received.

Because using glyphosate promotes the survival and reproductive success of GR weeds relative to non-GR weeds, using glyphosate has shortrun and longrun costs. The shortrun costs include material and application costs. The longrun costs include the eventual reduction of glyphosate effectiveness due to resistance, which leads to larger yield losses and higher costs of controlling GR weeds. Choices that manage resistance account for the relationship between glyphosate use and resistance over time, while choices that ignore resistance consider only the current growing season and not the temporal relationship between glyphosate use and resistance. As a result, choices to manage or ignore resistance lead to different results over time, including disparities in glyphosate resistance and weed densities, which in turn lead to differences in management costs, crop yields, and returns.

For illustrative purposes, we modeled one weed species—horseweed, one of the more widespread GR weed species affecting U.S. crop acreage—and three cropping scenarios: corn-soybean rotation, continuous corn, and continuous soybean.[15] We summarized returns, crop yields, herbicide costs, and weed densities for managing and ignoring resistance for a 20-year simulation period.

---

[15]We also examined managing and ignoring resistance choices for common waterhemp. Because the results for common waterhemp are very similar to those for horseweed, we discuss the latter only for simplicity. Palmer amaranth is another important weed species developing glyphosate resistance, particularly in Southeastern States. This report focuses on GR horseweed, because it was found earlier than palmer amaranth and in more States, including the Corn Belt (Boerboom and Owen, 2006).

Managing Glyphosate Resistance Differs from Ignoring Resistance

The simulation results show that managing resistance differs from ignoring resistance in three important ways. First, managing resistance entails using glyphosate during fewer years than ignoring resistance. Managing resistance would require use of glyphosate in 9, 6, and 9 of the 20 simulated years for the corn-soybean, continuous-corn, and continuous-soybean scenarios, respectively. Ignoring resistance requires use of glyphosate in 12, 7, and all 20 of the simulated years for those same scenarios.

Second, managing resistance would combine glyphosate with more herbicides than ignoring resistance. Managing resistance would combine glyphosate with both of the other herbicides included in this study in 7 of the 9 years glyphosate is used for the corn-soybean scenario, 6 of 6 years for continuous corn, and 7 out of 9 years for continuous soybean. Ignoring resistance would combine glyphosate with only the residual herbicide in 33 percent of the years glyphosate is used in soybean production for the corn-soybean scenario (2 of the 6 years) and 22 percent of the years glyphosate is used for the continuous-soybean scenario. Managing resistance always combines glyphosate with the other two herbicides whenever glyphosate is used in corn. Ignoring resistance generally combines glyphosate with the residual herbicide, but never adds an application of the alternative, post-emergence herbicide.

Third, and most importantly, managing resistance generally entails not applying glyphosate during consecutive growing seasons. When managing resistance, there is a larger interval between years glyphosate is used than when ignoring resistance.[16] Except for the continuous-soybean scenario, managing resistance alternates years when glyphosate is used with at least 1 year when glyphosate is not used during the initial, simulated growing seasons. Choices that ignore resistance involve using glyphosate during the first 5 consecutive years for each scenario, and each year for the continuous-soybean scenario. As a result, glyphosate resistance develops much more quickly when resistance is ignored than when it is managed.

Results from the bio-economic model show that managing resistance reduces returns in the first year of implementation but increases returns in the second and all subsequent years (fig. 4). The cumulative impact of the returns received, reported as the annualized present value (APV) of returns, when managing resistance exceeds that received when ignoring resistance after 2 consecutive years (fig. 4).[17] That is, managing glyphosate resistance will recover its initial lower APV of returns relative to ignoring resistance in about 2 years.[18]

In addition, the difference in returns between managing resistance and ignoring resistance increases with the number of years of consecutive use. The gap widens because horseweeds are controlled much more effectively, and crop yields are higher, when managing resistance. Herbicide costs are

---

[16]When managing resistance, the interval averages 1.4 years, 2.6 years, and 1.4 years between years when glyphosate is used in the corn-soybean, continuous-corn, and continuous-soybean scenarios, respectively. When ignoring resistance, the interval averages 0.6 years, 2.0 years, and 0 years for those same scenarios, respectively.

[17]The annualized present value (APV) of returns is the return that, if received each year, would equal the observed present value of returns received over a particular time horizon. See Appendix 1.

[18]Changes to the biological model that increase or decrease the rate that resistance develops can increase the APV of returns received ignoring resistance relative to managing resistance during the first couple of years.

13

The Economics of Glyphosate Resistance Management in Corn and Soybean Production, ERR-184
Economic Research Service/USDA

also generally slightly higher, but the value of yield gains always exceeds the herbicide cost after less than 2 years. These findings suggest that the longrun benefits of managing resistance far exceed the additional shortrun costs.

Figure 4

Differences between APV of returns received when managing resistance and when ignoring resistance, by cropping scenario

Dollars per acre per year

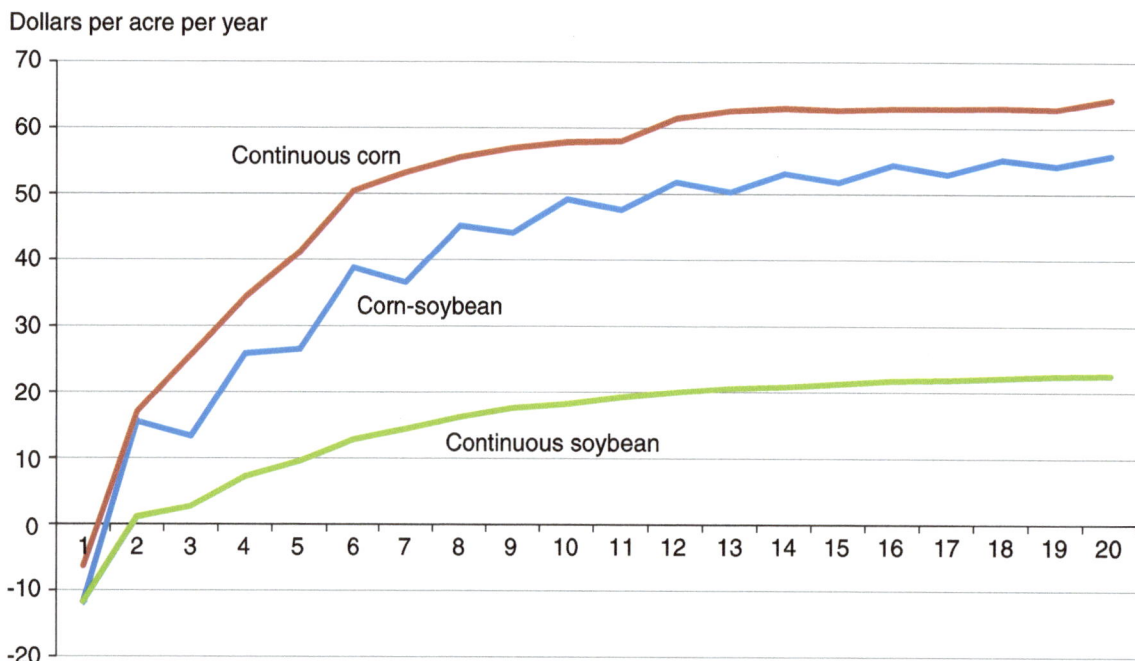

Years of consecutive use

Note: APV = annualized present value.
Source: USDA, Economic Research Service simulation results based on bio-economic optimization model; see also Appendix 1.

Impacts of Weed-Seed Dispersal

The simulation results suggest that much more weed seed is produced in fields where resistance is ignored (using a short-term perspective) than in fields where resistance is managed (using a longer term perspective) (table 2). This finding suggests that the longrun benefits of managing resistance might be lower in fields located near fields where resistance is ignored.

To explore this hypothesis, we modified the base model to simulate herbicide choices and returns to a "grower" for managing or ignoring resistance, for 3 crop rotations, when the "neighbor" either manages or ignores resistance (table 3).[19] The results show that for each crop rotation, the grower receives higher returns when: (1) the grower manages resistance, regardless of the neighbor's practice, and (2) the neighbor manages resistance, regardless of the grower's practice.

One implication is that the grower receives a gain from managing resistance relative to ignoring it, and that the gain is greater when the neighbor also manages resistance (thereby producing and dispersing fewer weed seeds). For the example of corn-soybean rotation, the grower's gain from managing resistance in his/her own field is $58.92 per acre when the neighbor manages resistance and $24.42 when the neighbor ignores resistance. However, for continuous soybeans, the gain from managing resistance is less than for the other rotations and is negligible when the neighbor

Table 2

Simulated APV of returns and herbicide costs, crop yields, and weed-seed densities for fields farmed managing resistance compared to fields farmed by ignoring resistance—by cropping scenario for a 20-year period

Item	Corn-soybean			Continuous-corn			Continuous-soybean		
	Managing resistance	Ignoring resistance	Difference	Managing resistance	Ignoring resistance	Difference	Managing resistance	Ignoring resistance	Difference
Annualized present value (2010 US$)									
Returns (per acre)	378.7	322.9	55.8	431.4	367.0	64.3	183.3	160.7	22.6
Herbicide cost per acre	25.0	20.4	4.5	22.7	20.5	2.2	27.4	20.3	7.1
Mean yield (bushels)									
Corn	202.0	182.2	19.9	189.8	176.2	13.5	NA	NA	NA
Soybean	58.2	55.8	2.4	NA	NA	NA	50.8	47.9	2.8
Mean weed-seed density (seeds per square meter)									
	11	1,615	1,604	11	438	427	11	4,164	4,154

Note: These results are based on the bioeconomic optimization model described in appendix 1. These results are for a 20-year planning horizon. NA = Not applicable.

Source: USDA, Economic Research Service using data from the Agricultural Resource Management Survey (ARMS) Phase II.

[19]In the base model, one seed per square meter is assumed to land on the representative crop field from an external source each year to simulate the impact of nearby, wild horseweed populations that are not selected for glyphosate resistance. See Appendix 1—Bio-economic Optimization Model for more details.

15

The Economics of Glyphosate Resistance Management in Corn and Soybean Production, ERR-184
Economic Research Service/USDA

Table 3

Simulated returns (APV) for a corn/soybean grower given the grower's management decisions and those of the grower's neighbor, three cropping scenarios

Corn-soybean rotation		Neighbor		
		Manages resistance	Ignores resistance	Grower's gain from neighbor managing resistance
		Dollars per acre		
Grower	Manages resistance	378.36	336.19	42.16
	Ignores resistance	319.44	311.77	7.67
	Grower gain from managing resistance in own field	58.92	24.42	NA

Continuous corn		Neighbor		
		Manages resistance	Ignores resistance	Grower's gain from neighbor managing resistance
		Dollars per acre		
Grower	Manages resistance	431.72	410.07	21.65
	Ignores resistance	368.19	360.47	7.72
	Grower gain from managing resistance in own field	63.53	49.60	NA

Continuous soybeans		Neighbor		
		Manages resistance	Ignores resistance	Grower's gain from neighbor managing resistance
		Dollars per acre		
Grower	Manages resistance	182.70	154.85	27.85
	Ignores resistance	158.79	154.82	3.97
	Grower gain from managing resistance in own field	23.91	0.04	NA

Notes: Results from the bio-economic optimization model described in appendix 1. These results (APV) are for a 20-year planning horizon with a discount factor per year equal to 0.95 (Lence, 2000). APV = annualized present value. NA = Not applicable.

Source: USDA, Economic Research Service using data from the Agricultural Resource Management Survey (ARMS) Phase II.

ignores resistance, only $0.04 per acre.[20] The grower has a larger gain when the neighbor manages resistance because the simulated weed-seed dispersal has less impact on the grower's returns than when the neighbor ignores resistance. Therefore, one effect of the neighbor's ignoring resistance is to reduce the grower's financial incentive to manage resistance. The difficulty in determining how operators of nearby farms manage weeds may likewise reduce incentive and, in turn, explain why some crop growers do not manage resistance, especially in soybean production.

---

[20]Also for continuous soybeans only, the grower's ignoring resistance, when the neighbor manages resistance, would have slightly greater returns ($158.79 per acre), than the grower's managing resistance when the neighbor ignores it ($154.85). But when the neighbor manages resistance, the grower achieves even greater returns by managing resistance ($182.70).

16

The Economics of Glyphosate Resistance Management in Corn and Soybean Production, ERR-184
Economic Research Service/USDA

Other implications of grower-neighbor interdependence are (1) that the grower has a financial incentive to encourage the neighbor to manage resistance because grower returns are greater when the neighbor manages rather than ignores resistance and (2) that the incentive is greater when the grower also manages resistance. For example, for the corn-soybean rotation, the grower managing resistance in his/her own field gains $42.16 per acre from the neighbor managing resistance, but the grower ignoring resistance gains only $7.67 from the neighbor's managing resistance.

These results also suggest that the benefit of free-riding on the resistance management efforts of neighboring growers is relatively small. The returns received when the grower ignores resistance can remain low even if a neighbor manages resistance, and the grower's gain from the neighbor's managing resistance is also relatively small (for example, $3.97 per acre for continuous soybeans). The benefit that a grower's managing resistance transfers to neighbors is fewer and less frequent immigrant, GR weed seeds. Because this benefit is difficult to observe, it is easy to ignore, which suggests that improving communications and teamwork between neighboring growers could help promote resistance management.

Weed Susceptibility to Herbicides Is a Common Pool Resource

The simulation results show that the grower's gain from managing resistance is lower when the neighbor ignores resistance (table 3), which suggests that the common-pool-resource nature of weed susceptibility to glyphosate can reduce the returns possible from the grower's managing resistance (long-term perspective) relative to the grower's ignoring resistance (short-term view). For two reasons, common pool resources create a potential social dilemma, in which users seeking short-run, personal objectives lead to outcomes that are not desirable from anyone's longrun perspective: (1) it is particularly costly to exclude those growers who have a shortrun view from using the resource (non-excludability), and (2) their use by each grower reduces its availability to others (subtractability) (Ostrom et al., 1999). Without effective rules limiting access and providing incentives for users to invest in rather than exploit the resource, common pool resources are vulnerable to overuse. In our case, the common-pool-resource nature of weed susceptibility to glyphosate can reduce economic incentives to manage resistance. As a result, some crop growers might opt instead for short-term herbicide choices, hastening glyphosate resistance.

Ostrom et al. (1999) document a wide range of common pool resources that have been effectively managed by resource users without government intervention, especially when people are able to communicate, sanction each other, or make new rules. As long as the proportion of free riders is not too high initially, cooperative agreements can be established, sustained, and expanded by resource users. Thus, the simulation results suggest that returns to glyphosate-resistance management can be increased by cooperation between crop growers. In addition, as we describe below, recent survey data suggest that the proportion of free-riders is not too high to preclude cooperation, which has already occurred in Arkansas (Smith, 2012) and more recently in North Carolina (Everman, 2014).

To recap, the simulations suggest that herbicide choices that manage resistance pursuing a longrun economic goal result in higher returns over time than do herbicide choices that ignore resistance and pursue a shortrun economic goal. Managing resistance uses less glyphosate, combines it with alternative herbicides, and generally does not use it during consecutive growing seasons, so that resistance develops less quickly and returns are higher. However, the simulations also show that the common-pool-resource nature of weed susceptibility to glyphosate can reduce economic incentives for crop growers to use resistance-managing herbicide choices.

The Economics of Glyphosate Resistance Management in Corn and Soybean Production, ERR-184
Economic Research Service/USDA

Glyphosate-Resistant Weeds and Corn and Soybean Grower Responses

To better understand the trends in weed-management practices used in corn and soybean production since the commercial availability of GT varieties, it is necessary to determine how many corn and soybean acres are currently affected by GR weeds and how corn and soybean growers are responding. As already shown, glyphosate use increased steadily on corn and soybean planted acres, as did the number of acres that received glyphosate by itself. However, sometime between 2006 and 2012 in soybean production, and 2005 and 2010 in corn production, the number of acres receiving glyphosate by itself declined as the number receiving glyphosate plus at least one additional herbicide MOA increased. As we will show, this reversal was associated with management responses to GR weeds in both crops.[21]

The 2010 Phase ARMS survey asked corn growers, "Has [the surveyed] field ever been infested with weeds resistant to glyphosate?" According to the growers' responses, GR-weed infestations had been reported in only 5.6 percent of planted corn acres (4.5 million acres) at that time. The majority of those acres were in the Corn Belt (34 percent) and Northern Plains (46 percent), which account for the majority of U.S. corn acres. GR-weed infestations were reported in less than 4 percent and 10 percent of corn acres in those regions, respectively. GR-weed infestations were reported in higher percentages of corn acres in southern regions; over a quarter of corn acres planted in Georgia in 2010 had been infested with GR weeds as of 2010.

The 2012 Phase-II ARMS survey asked soybean growers, "Have you noticed a decline in the effectiveness of glyphosate in controlling weeds in [the surveyed] field?" The growers reported a decline in glyphosate's effectiveness in 43.7 percent of planted soybean acres (32.5 million acres). Again, the majority of those acres were in the Corn Belt (47 percent) and Northern Plains (23 percent), which also typically account for the majority of U.S. soybean acres. A decline in the efficacy of glyphosate was reported in higher percentages of soybean acres in southern regions, including Appalachia (58 percent) and the Delta (55 percent), but the Corn Belt (46 percent), Northern Plains (40 percent), and Lake States (31 percent) were not far behind.

Responses to these questions suggest that both corn and soybean growers believed they had experienced problems with GR weeds by 2010 and 2012, respectively, and that more soybean acres in 2012 than corn acres in 2010 were affected. However, because somewhat different questions were asked in 2010 and 2012, we cannot definitively conclude that the prevalence of GR weeds differs in corn and soybean fields.

Corn and soybean growers responded similarly to their respective issues of the presence of GR weeds and decline in glyphosate effectiveness. The most popular response was to use herbicides other than glyphosate on over 84 percent of corn acres with GR weeds and on 71 percent of soybean acres where declines in glyphosate effectiveness were reported. Increasing the amount of glyphosate used was the next most popular response, with a reported 25 percent of corn with GR weeds receiving additional glyphosate and 39 percent of soybean acres with declines in glyphosate

---

[21]It is possible to see this reversal by considering growers' responses to questions related to corn in 2010 and soybeans in 2012.

19

The Economics of Glyphosate Resistance Management in Corn and Soybean Production, ERR-184
Economic Research Service/USDA

effectiveness receiving more glyphosate.[22] Corn and soybean growers changed their tillage practices on almost 7 percent and over 14 percent, respectively, of each crop's acreage reported to have GR weeds or declines in glyphosate effectiveness. Very few growers of corn (2.3 percent of affected acreage) and soybean (3.2 percent of affected acreage) reported reducing use of GT corn or soybean varieties in response to GR weeds or declining glyphosate effectiveness.

The result of the simulations suggest that increasing the use of other herbicide MOAs can help delay glyphosate resistance, especially when glyphosate is not used or, if used, is combined with at least one different herbicide MOA. Because tillage controls weeds without promoting glyphosate resistance, increasing the intensity of tillage can also help delay resistance and reduce the spread of GR weeds. Planting fewer acres to GT crops or planting GT varieties tolerant to other herbicides can also help, if operators also reduce glyphosate use. However, increasing the amount of glyphosate used can increase the rate of glyphosate resistance.

Consistent with the simulations and with plant-scientists' recommendations, herbicide-use practices on over 82 percent of corn acres in 2010 were a substantial component of glyphosate-resistance management. Growers either combined glyphosate with at least one different MOA, or else used herbicides with other MOAs and without glyphosate. Between acreage with and without reported GR weeds, no statistically significant difference exists in the prevalence of these herbicide-use practices. Glyphosate was used by itself on almost 18 percent of planted acres.[23]

Herbicide-use practices on 60 percent of soybean acres in 2012 were a substantial component of glyphosate resistance management. However, there are statistically significant differences in the prevalence of resistance management practices between soybean acres with and without reported declines in glyphosate effectiveness. Herbicide-use practices were consistent with resistance management on more soybean acres with reported declines in glyphosate effectiveness (66 percent) than without such reported declines (50 percent). For example, more soybean acres where growers reported such declines received glyphosate and at least one herbicide with a different MOA (62 percent) than did acres where growers did not report declines (45 percent). In addition, fewer soybean acres with reported declines in glyphosate effectiveness (31 percent) than without such declines (46 percent) received glyphosate alone.

These findings suggest that glyphosate-resistance management could have been improved on more soybean acres in 2012 (up to 40 percent) than on corn acres in 2010 (up to 18 percent), by applying glyphosate with at least one other herbicide with a different MOA, or by applying an alternative herbicide instead of glyphosate. Moreover, these findings suggest that herbicide choices that delay resistance were more likely to be used proactively in corn than soybean production and that such herbicide choices in soybean production were more likely to be reactive—that is, in response to glyphosate resistance.

---

[22]On over 57 percent of acres with GR weeds reported, corn growers did not change their use of glyphosate; on 5 percent of acres, they used less; and on 10 percent, they stopped using it. On over 39 percent of acres with reported declines in glyphosate effectiveness, soybean growers did not change their use of glyphosate; on 9 percent of such acres, they used less; and on a little over 2 percent, they stopped using it.

[23]Glyphosate was used by itself on corn in each of the major production regions, but was most prevalent in the Southern Plains, followed by the Southeast, Lake States, and Northern Plains regions.

Economic Impacts of Glyphosate Resistance as of 2010 and 2012

The simulation results suggest that glyphosate resistance can be costly and that it can be economically beneficial to manage instead of ignore resistance (see table 3). Because the simulation results are based on a simplified model of reality, we complement the simulation results with more direct estimates of the costs of resistance. ARMS data provide detailed information on production costs, yields, and returns that can be used in conjunction with responses to the glyphosate resistance/effectiveness questions to estimate the economic impacts of GR weeds on both corn and soybean production.

Estimating the impact of being in one group (for example, corn growers who reported a GR-weed infestation—the treatment group) instead of another group (corn growers who did not report a GR-weed infestation—the control group) by comparing the group means for an outcome variable of interest (e.g., returns per acre) can be misleading because of sample-selection bias (Heckman, 1979). The characteristics of growers and their farms can influence both the group they are in and their economic performance.

For example, growers who operated smaller corn enterprises in 2010 were more likely to report GR-weed infestations than did growers who operated larger corn enterprises, which are often more profitable than smaller ones. Without accounting for the influence of enterprise size, comparing average returns per acre for growers who reported a GR-weed infestation to returns for those who did not, would likely provide a biased estimate of the economic impact of glyphosate resistance.

Statisticians (Rosenbaum and Rubin, 1983), economists (Abadie and Imbens, 2006), and political scientists (Diamond and Sekhon, 2013) have shown that grouping very similar farms together before comparing mean outcomes can reduce sample-selection bias. The procedure, known as propensity-score matching, first chooses treatment and control groups (e.g., growers with and without reported GR weeds, respectively) that are statistically similar, based on a wide variety of characteristics other than the presence of GR weeds (e.g., enterprise size), before comparing group means (e.g., of per-acre returns). (See appendix 2 for more details.)

We used a propensity-score matching procedure to estimate the impacts of glyphosate-resistance on corn and soybean production in 2010 and 2012. The results suggest that corn growers who had reported a GR-weed infestation in 2010 realized significantly lower operating (-$60.19/acre) and total (-$67.29/acre) returns than similar corn growers who had not reported such an infestation (table 4). The results suggest that lower yields and higher chemical and fuel costs might have contributed to the shortfall in returns, although the differences in yields and chemical and fuel costs themselves were not statistically significant at the 10-percent level. These findings suggest glyphosate resistance contributed to a substantial reduction in returns to affected corn growers.

Similarly, a propensity-score matching procedure suggests that soybean growers who had reported a decline in the effectiveness of glyphosate as of 2012 received lower total ($22.53/acre) returns than soybean growers with similar characteristics who had not reported such a decline (table 5). The estimates suggest that lower yields and higher chemical costs might have contributed to the lower returns, although the difference in yields is not statistically significant at the 10-percent level. These findings suggest that glyphosate resistance contributed to a 14-percent reduction in returns to affected soybean growers.

21

The Economics of Glyphosate Resistance Management in Corn and Soybean Production, ERR-184
Economic Research Service/USDA

Table 4

Impacts on returns, yield, and production costs of reporting a glyphosate-resistant weed infestation in corn, 2010

Outcome variable	Reported a GR-weed infestation		Impact	Standard error
	Yes	No		
Yield (bushels/ harvested acre)	133.74	143.29	-9.54	6.64
Total production costs ($/planted acre)	558.37	538.49	19.88	19.36
Operating costs ($/planted acre)	295.99	283.21	12.78	12.73
Allocated overhead costs ($/planted acre)	262.38	255.27	7.11	11.90
Total returns ($/planted acre)	35.10	102.39	-67.29**	32.31
Operating returns ($/planted acre)	297.48	357.66	-60.19**	29.63

Notes: Propensity-score-matching estimates based on 2010 Phase II and Phase III ARMS data. 1,607 observations are in the initial dataset; 95 observations are in both matched samples. Matching statistics are reported in appendix 2.
*, **, and *** indicates statistical significance at 10-, 5-, and 1-percent levels, respectively. GR = glyphosate resistant.

Source: USDA, Economic Research Service using data from the 2010 Agricultural Resource Management Survey (ARMS) Phase II.

Table 5

Impacts on returns, yield, and production costs of having reported a decline in glyphosate effectiveness in controlling weeds in soybean, 2012

Outcome variable	Observed decline in effectiveness of glyphosate in controlling weeds		Impact	Standard error
	Yes	No		
Yield (bushels/ harvested acre)	41.08	42.16	-1.08	1.00
Total production costs ($/planted acre)	445.41	437.76	7.66	8.27
Operating costs ($/planted acre)	184.23	178.97	5.27	4.81
Overhead costs ($/planted acre)	261.18	258.79	2.39	5.76
Total returns ($/planted acre)	100.80	162.36	-22.53*	13.46
Operating returns ($/planted acre)	401.01	421.15	-20.14	13.66

Notes: Propensity-score-matching estimates based on 2012 Phase II and Phase III ARMS data. The initial dataset contains 1,856 observations; both matched samples contain a total of 791 observations. Matching statistics are reported in appendix 2. *, **, and *** indicates statistical significance at 10-, 5-, and 1-percent levels, respectively.

Source: USDA, Economic Research Service using data from the 2012 Agricultural Resource Management Survey (ARMS) Phase II.

Economic Impacts of Using Glyphosate by Itself in Corn and Soybean Production

Because both the literature and our simulation results suggest that using glyphosate by itself is the main factor underlying the evolution of glyphosate resistance, estimating the impacts on production costs, yields, and returns of using this practice is critical. If the returns when using glyphosate by itself exceeded, in the shortrun, returns when using herbicides that delay resistance, there would be an important barrier to resistance management.

We used propensity-score matching to estimate the economic impacts of using glyphosate by itself, compared to using glyphosate and at least one different herbicide MOA, in corn and soybean production. The estimates suggest that corn growers who used glyphosate and at least one different herbicide MOA had higher operating costs ($15.36/acre due in part to higher chemical costs), yields, returns, and operating returns than did corn growers who were very similar, but used glyphosate by itself. However, differences in yields and returns were not statistically significant at the 10-percent level (table 6).

The estimates suggest that soybean growers using glyphosate and at least one different herbicide MOA received statistically higher yields (4.31 bushels/acre), returns ($48.57/acre), and operating returns ($49.23/acre)—despite having higher operating costs ($11.41/acre)—than similar soybean growers using glyphosate by itself (table 7).

Our findings for corn, but not for soybeans, are broadly consistent with results from Weirich et al. (2011), who used Benchmark Study data for the period 2006-2008 to study corn and soybean growers from Iowa, Illinois, Indiana, Mississippi, Nebraska, and North Carolina. Weirich et al. (2011) compared the costs, yields, and returns from using the growers' usual weed control methods

Table 6

Impacts on returns, yield, and production costs of using glyphosate by itself rather than using glyphosate and at least one different herbicide in corn, 2010

| | Used glyphosate | | | |
Outcome variable	By itself	With one or more other herbicides	Impact	Standard error
Yield (bushels/harvested acre)	133.94	139.88	-5.94	4.20
Total production costs ($/planted acre)	511.27	531.65	-20.38	13.88
Operating costs	255.31	270.67	-15.36**	7.71
Overhead costs	255.95	260.98	-5.02	10.17
Total returns ($/planted acre)	64.79	80.51	-15.72	20.81
Operating returns	320.75	341.49	-20.74	17.72

Notes: Propensity-score-matching estimates based on 2010 Phase II and Phase III ARMS data. 963 observations are in the initial dataset; 343 observations are in both matched samples. Matching statistics are reported in appendix 1. Matching statistics are reported in appendix 1.
*, **, and *** indicates statistical significance at 10-, 5-, and 1-percent levels, respectively.

Source: USDA, Economic Research Service using data from the 2010 Agricultural Resource Management Survey (ARMS) Phase II.

23

The Economics of Glyphosate Resistance Management in Corn and Soybean Production, ERR-184
Economic Research Service/USDA

Table 7

Impacts on returns, yield, and production costs of using glyphosate by itself rather than using glyphosate and at least one different herbicide in soybean, 2012

| Outcome variable | Used glyphosate | | Impact | Standard error |
	By itself	With one or more other herbicides		
Yield (bushels/harvested acre)	38.93	43.23	-4.31***	0.98
Total production costs ($/planted acre)	425.53	437.61	-12.08	7.49
Operating costs	173.42	184.83	-11.41**	4.68
Overhead costs	252.11	252.78	-0.67	5.47
Total returns ($/planted acre)	126.41	174.98	-48.57***	13.46
Operating returns	378.52	427.76	-49.23***	13.91

Notes: Propensity-score-matching estimates based on 2012 Phase II and Phase III ARMS data. 1,710 observations are in the initial dataset; 753 observations are in both matched samples. Matching statistics are reported in appendix 1.
*, **, and *** indicates statistical significance at 10-, 5-, and 1-percent levels, respectively.

Source: USDA, Economic Research Service using data from the 2012 Agricultural Resource Management Survey (ARMS) Phase II.

(e.g., using glyphosate) with the same measures resulting from following the academics' recommendations (e.g., use other herbicides in combination with glyphosate). They conclude that weed management costs were higher with more intensive management with herbicides. However, reduced weed pressure resulted in a trend toward higher crop yields, which offset the higher weed management costs. They also observe that the grower's returns "will be equivalent in the short term, and, over time, long-term resistance management will delay the evolution of GR weeds in their fields, creating substantial additional saving." Similarly, our shortrun results for corn in 2010 indicate that the differences in mean yields and returns were not statistically significant between fields where glyphosate was applied with at least one different herbicide MOA and fields where glyphosate was used by itself (see table 6).

However, our results for soybeans in 2012 show that the mean yields and returns were higher in fields where glyphosate was applied with at least one different herbicide MOA than in fields where glyphosate was used by itself (see table 7). The findings for soybeans suggest that using glyphosate by itself may not only exacerbate glyphosate resistance, but could also be a costly short-term management decision. Currently, shortrun economic disincentives do not appear to impede the use of glyphosate with at least one different herbicide MOA (at least for soybeans).

The simulation results from the bio-economic model show that the returns when managing resistance are higher in the long term but lower in the first 1-2 years (fig. 4) than when ignoring resistance. Thus, taking all the results together, we conclude that managing glyphosate resistance produces higher returns than does ignoring resistance, except possibly the first 2 years.

Soybean Growers' Beliefs About the Common-Pool-Resource Nature of Weed Susceptibility to Glyphosate

One of the key findings from the simulation analysis is that weed-seed dispersal between farms can reduce the economic incentives for managing resistance. Because weed susceptibility to glyphosate is an example of a common pool resource, it can encourage glyphosate overuse, which can hasten glyphosate resistance. However, a wide range of common pool resources have been effectively managed by resource users who were able to communicate, sanction each other, or make new rules in situations where the proportion of free-riders was not too high (Ostrom et al., 1999). Under these circumstances, cooperative agreements have been established, sustained, and expanded by resource users.

We used ARMS (phase 2) data collected from soybean growers in 2012 to identify the use of any of seven commonly recommended RMPs ever used to delay glyphosate resistance, to indicate whether weed-management costs increased as a result of using these practices, and to summarize perceptions on whether or not the RMPs would be more effective in delaying herbicide resistance if nearby growers also used them.[24] Crop rotation to delay glyphosate resistance was used on the largest percentage of GT soybean acreage (79 percent), followed by using the herbicide-label application rate (70 percent), controlling weeds at the appropriate time in their growth stage (70 percent), using herbicides other than glyphosate (59 percent), using tillage (42 percent), ensuring all weeds are killed after herbicide use (41 percent), and cleaning equipment between fields (24 percent).

Weed-management costs increased on 20 percent of RMP acres using crop rotation, 32 percent using the herbicide-label-recommended application rate, 35 percent controlling weeds at the appropriate time, 62 percent using herbicides other than glyphosate, 41 percent using tillage, 53 percent ensuring all weeds are killed after herbicide use, and 21 percent cleaning equipment between fields. These responses suggest that cost influenced the extent of RMP use.

To investigate whether the common-pool-resource nature of weed susceptibility to glyphosate influenced the use of RMPs, in the 2012 soybean ARMS survey we also asked growers if they believed the RMPs they used would be more effective in delaying herbicide resistance if operators of nearby farms also used them (the common-pool-resource beliefs question). Almost 53 percent of the soybean acres that used at least one RMP were farmed by soybean growers who answered "yes"; another 20 percent answered "no" (fig. 5).[25] Smaller shares of acres "with GR weeds" (reporting a decline in the effectiveness of glyphosate in controlling weeds) (20 percent) than without (28 percent) were managed by soybean growers who answered "don't know," and these estimates are statistically different.[26]

---

[24]The question was "Consider each year you planted a GR crop in this field, have you ever used the following practices in order to reduce the rate that glyphosate resistance develops in weeds on this field?" Seven resistance management practices were included. For each of the practices, the questionnaire asked "How often did you use this practice on this field?" and also "Did the cost of managing weeds on this field increase as a result of your use of the practice?"

[25] Beliefs about the common-pool-resource nature of weed susceptibility to glyphosate varied regionally, with higher acreage shares operated by soybean growers who answered "yes" located in the Appalachia (70 percent) and Lake States (58 percent) regions; those answering "no" were more often located in the Corn Belt (23 percent) and Northern Plains (22 percent) regions.

[26]More soybean acres "with GR weeds" (55 percent) than "without GR weeds" (51 percent) were farmed by soybean growers who answered "yes," although the difference in these estimates is not statistically significant. The difference in the shares of soybean acres with and "without GR weeds" farmed by soybean growers who answered "no" or "the nearest farm is too far away..." also is not statistically significant.

25

The Economics of Glyphosate Resistance Management in Corn and Soybean Production, ERR-184
Economic Research Service/USDA

Figure 5

Responses of soybean growers with and without GR weeds to the common-pool-resource belief question in ARMS, by percent of soybean acres, 2012

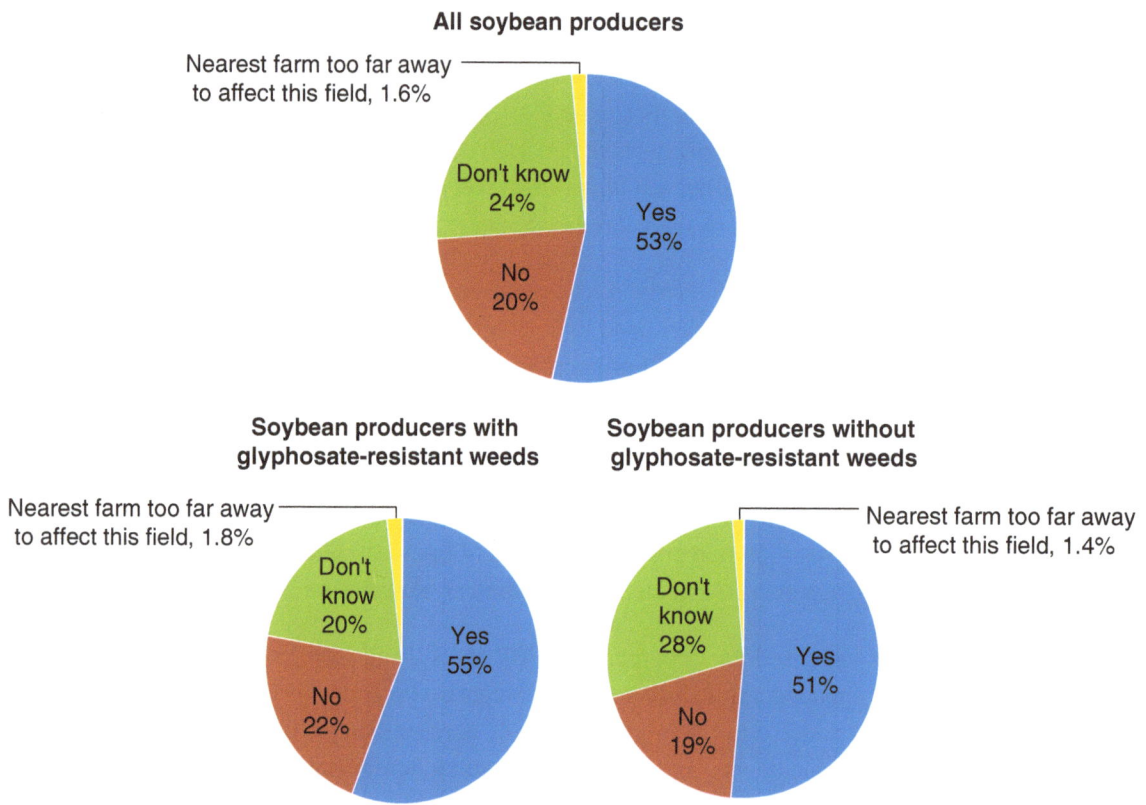

Notes: Growers' answers are in response to the question "Do you believe the resistance-management practices you used are or would be more effective in reducing the rate that herbicide resistance develops in weeds on your farm if operators of nearby farms also used them?"
Source: USDA, Economic Research Service using data from the 2012 Agricultural Resource Management Survey (ARMS) Phase II.

Sample averages indicate that soybean growers who answered "yes" to the common-pool-resource beliefs question used more RMPs than did other soybean growers; soybean growers who answered "don't know" used fewer RMPs (tables 8 and 9 and Appendix 2). The number of RMPs used by soybean growers who answered "no" or "the nearest farm is too far away…" are not statistically significantly different from the number of RMPs used by other growers.

We used propensity-score matching to estimate differences in the numbers of RMPs used, as well as differences in yields, production costs, and returns among growers holding different beliefs about the common-pool-resource nature of weed susceptibility to glyphosate. Soybean growers who answered "yes" to the common-pool-resource beliefs question used 17 percent more RMPs (p<0.01) than did similar soybean growers who chose another answer (table 8). The "yes" respondents also received higher yields (2.18 bushels/acre), higher returns ($35.62/acre), and higher operating returns ($28.98/acre).

Soybean growers who answered "yes" were more likely to use herbicides other than glyphosate, use the herbicide-label-recommended application rate, control weeds at the appropriate time in their life stage, and ensure that all weeds were killed after herbicide use than soybean growers who answered the common-pool-resource beliefs question differently. Although crops were rotated on over 85

Table 8

2012 outcomes for soybean growers who answered "yes" to the common-pool-resource belief question in ARMS

Outcome variable	Believed RMPs more effective if neighbors also used them		Impact	Standard error
	Yes	All other answers		
Number of RMPs used	4.49	3.85	0.64***	0.11
Yield (bushels/ harvested acre)	42.56	40.38	2.18**	0.92
Total production costs ($/planted acre)	441.61	445.67	-4.06	8.01
Operating costs ($/planted acre)	184.15	181.56	2.58	4.67
Overhead costs ($/planted acre)	257.46	264.11	-6.64	5.95
Total returns ($/planted acre)	163.90	128.28	35.62***	12.73
Operating returns ($/planted acre)	421.36	392.39	28.98***	12.94

Notes: Propensity-score-matching estimates based on 2012 Phase II and Phase III ARMS data. 1555 observations are in the initial dataset; 803 observations are in both matched samples. Matching statistics are reported in Appendix 2. *, **, and *** indicates statistical significance at 10-, 5-, and 1-percent levels, respectively. RMP = resistance management practice. The common-pool-resource belief question was "Do you believe the resistance-management practices you used are or would be more effective in reducing the rate that herbicide resistance develops in weeds on your farm if operators of nearby farms also used them?"

Source: USDA, Economic Research Service using data from the Agricultural Resource Management Survey (ARMS) Phase II.

percent of the soybean acres managed by growers who answered "yes," this practice was used to a similar extent by soybean growers who answered differently, perhaps because rotation has benefits unrelated to the spread of GR weeds and other weed control considerations.

Soybean growers who answered "don't know" used 7 percent fewer RMPs than other, very similar soybean growers who chose one of the other answers (table 9). Soybean growers who were uncertain about the common-pool-resource nature of weed susceptibility to glyphosate received lower yields (-2.20 bushels per acre) and lower total (-$46.01/acre) and operating (-$39.00/acre) returns.

Because it was difficult to determine whether or not nearby growers used RMPs, and which ones they used, it was not clear whether growers who answered "yes" to the common-pool-resource beliefs question would use fewer or more RMPs. Conceivably, growers who answered "yes" might use fewer RMPs if they expected to receive GR weeds from their neighbors, which would undermine efforts to delay glyphosate resistance. However, instead, growers who answered "yes" routinely reported to have used more RMPs and performed better financially than soybean growers who either did not believe in, or were uncertain about, the common-pool-resource nature of weed susceptibility to glyphosate.

The findings in table 8, viewed in combination with our simulation results on the impacts of weed-seed dispersal (see table 3), suggest that soybean growers aware of the common-pool-resource nature of weed susceptibility to glyphosate not only had an economic incentive to convince neigh-

Table 9

2012 outcomes for soybean growers who answered "don't know" to the common-pool-resource belief question in ARMS

Outcome variable	Believed RMPs more effective if neighbors also used them		Impact	Standard error
	Uncertain	All other answers		
Number of RMPs used	3.74	4.01	-0.27**	0.13
Yield (bushels/ harvested acre)	39.65	41.85	-2.20**	1.04
Total production costs ($/planted acre)	451.18	436.62	14.56*	8.13
Operating costs ($/planted acre)	186.90	179.35	7.54	4.94
Overhead costs ($/planted acre)	264.28	257.26	7.02	5.82
Total returns ($/planted acre)	112.57	158.59	-46.01***	15.08
Operating returns ($/planted acre)	376.85	415.85	-39.00***	14.94

Notes: Propensity-score-matching estimates based on 2012 Phase II and Phase III ARMS data. 1,555 observations are in the initial dataset; 399 observations are in both matched samples. Matching statistics are reported in Appendix 2.
*, **, and *** indicates statistical significance at 10-, 5-, and 1-percent levels, respectively. The common-pool-resource belief question was "Do you believe the resistance-management practices you used are or would be more effective in reducing the rate that herbicide resistance develops in weeds on your farm if operators of nearby farms also used them?" RMP = resistance management practice.

Source: USDA, Economic Research Service using data from the Agricultural Resource Management Survey (ARMS) Phase II.

bors to use RMPs, but also might have been aware of the incentive to cooperate with their neighbors in managing glyphosate resistance.

Examples of cooperation among growers suggest that some crop growers recognize cooperation's benefits for managing glyphosate resistance. Glyphosate resistance has evolved in pigweed (Palmer amaranth) populations in the southern United States, where it infests cotton and soybean fields, reduces crop yields, and increases weed-management costs. Arkansas extension agents encouraged growers to use a zero-tolerance approach developed by University of Arkansas scientists to eliminate production of pigweed seeds in their crop fields by creating zero-tolerance zones and managing weed seed beds in crop fields, ditches, and turn rows with pre-plant and pre-emergence herbicides and tillage, when necessary (Smith, 2012; Yancy, 2012). In 2011, growers in Clay County, Arkansas, adopted this approach, followed by growers in Desha County in 2012. Also, soybean growers in North Carolina recently started a similar program that has been promoted aggressively (Everman, 2014).

Finally, recall that resistance-management practices, at least those used in 2012 soybean production, were used on higher percentages of acres with reported declines in glyphosate effectiveness than acres without such declines, and the differences are statistically significant in most cases. This result suggests that resistance-management choices in soybeans in 2012 were more likely to be used to react to glyphosate resistance, rather than to proactively delay resistance, because growers were more likely to use RMPs after, rather than before, they noticed declines in glyphosate effectiveness.

Conclusion

The findings of this study suggest that communications about (1) the negative economic consequences of glyphosate resistance, (2) the economic benefits of managing resistance instead of ignoring it, (3) the common-pool-resource nature of weed susceptibility to glyphosate, and (4) the potential benefits of cooperation among neighboring crop growers could help promote the use of resistance management practices (RMPs) and thereby increase the long-term returns to corn and soybean production.

While there are examples of neighboring growers in Arkansas and North Carolina cooperating to manage glyphosate-resistant weeds, further research is needed to determine the extent of such arrangements. Likewise, deserving further study is the role of outside stakeholder groups, such as agricultural extension specialists, crop consultants and university, industry, and government representatives, in the process of initiating, sustaining, and expanding these types of agreements.

For example, simply knowing about the common-pool-resource nature of weed susceptibility to glyphosate does not appear to be enough to ensure that resistance-managing herbicide choices are used proactively to delay resistance, rather than reactively to stem declines in glyphosate effectiveness. Our results suggest that further cooperation among growers, particularly soybean growers, may significantly reduce the rate of glyphosate resistance and increase grower returns.

Finally, economic research using future ARMS surveys and other data sources could continue to examine growers' use of weed-management practices, including strategies to manage weeds resistant to glyphosate or other herbicides and their effectiveness, and their effect on grower returns for corn, soybeans, and other crops such as cotton and wheat. Further studies may focus on (1) the rate and impacts of adoption of various HT seed varieties, including new ones tolerant to herbicides other than glyphosate, such as glufosinate with 2,4-D or dicamba; (2) changes in the use of herbicides, including glyphosate and other materials with various modes of action; (3) changes in the use of non-pesticide RMPs, such as increased tillage; and (4) the types and extent of cooperative arrangements among growers to manage resistance. In some cases, these studies may incorporate the potential economic and environmental tradeoffs between weed management and soil health due to the effects of tillage practices on both.

29

The Economics of Glyphosate Resistance Management in Corn and Soybean Production, ERR-184
Economic Research Service/USDA

References

Abadie, A., and G.W. Imbens. 2006. "Large Sample Properties of Matching Estimators for Average Treatment Effects," *Econometrica* 74(1):235-267.

Bellman, R. 1957. *Dynamic Programming.* Princeton University Press: Princeton, NJ.

Boerboom, C., and M. Owen. December 2006. *Facts About Glyphosate-Resistant Weeds.* Purdue University, GWC-1.

Chambers, R.G., and R. Lichtenberg. 1994. "Simple Econometrics of Pesticide Productivity," *American Journal of Agricultural Economics* 76(3):407-417.

Culpepper, A.S., J.R. Whitaker, A.W. MacRae, and A.C. York. 2008. "Distribution of glyphosate-resistant Palmer amaranth (Amaranthus palmeri) in Georgia and North Carolina during 2005 and 2006," *Journal of Cotton Science* 12:306-310.

Culpepper, A.S., and J. Kichler. 2009. *University of Georgia programs for controlling glyphosate-resistant Palmer amaranth in 2009 cotton.* College of Agricultural and Environmental Sciences, Circ. No. 924. University of Georgia, Athens, GA.

Dauer, J.T., E.C. Luschei, and D.A. Mortensen. 2009. "Effects of Landscape Composition on Spread of an Herbicide-Resistant Weed," *Landscape Ecology* 24:735-747.

Davis, V.M., and W.G. Johnson. 2008. "Glyphosate-Resistant Horseweed (*Conyza canadensis*) Emergence, Survival, and Fecundity in No-Till Soybean," *Weed Science* 56:231-236.

Diamond, A., and J.S. Sekhon. 2013. "Genetic Matching for Estimating Causal Effects: A General Multivariate Matching Method for Achieving Balance in Observational Studies," *Review of Economics and Statistics* 95(3):932-945.

Duke, S.O., and S.B. Powles. 2008. "Glyphosate: A Once-In-A-Century Herbicide," *Pest Management Science* 64(4):319–325.

Duke, S.O., and S.B. Powles. 2009. "Glyphosate-resistant crops and weeds: Now and in the future." *AgBioForum* 12(3&4):346-357.

Everman, W. June 2, 2014. Cooperation agreements among growers. In North Carolina State University, Department of Crop Science, personal communication to M. Livingston.

Farm Industry News. 2014. "Valent, Monsanto expand soybean cash-back program." http://farmindustrynews.com/herbicides/valent-monsanto-expand-soybean-cash-back-program

Fawcett, R., and D. Towery. 2002. *Conservation and Plant Biotechnology: How New Technologies Can Improve the Environment by Reducing the Need to Plow.* Conservation Technology Information Center, West Lafayette, IN.

Feder, G., and U. Regev. 1975. "Biological Interactions and Environmental Effects in the Economics of Pest Control," *Journal of Environmental Economics and Management* 2:75-91.

30

The Economics of Glyphosate Resistance Management in Corn and Soybean Production, ERR-184
Economic Research Service/USDA

Fernandez-Cornejo, J., R. Nehring, C. Osteen, S. Wechsler, A. Martin, and A. Vialou. 2014. *Pesticide Use in U.S. Agriculture, 1960-2008,* EIB-124. U.S. Department of Agriculture, Economic Research Service.

Fernandez-Cornejo, J., S. Wechsler, M. Livingston, L. Mitchell. 2014a. *Genetically Engineered Crops in the United States,* ERR-162. U.S. Department of Agriculture, Economic Research Service.

Florence, P.S. 1933. *The Logic of Industrial Organization,* P. Kegan, editor. Trench & Trubner, London.

Fraser. K. January 25, 2013. *Glyphosate Resistant Weeds – Intensifying,* Stratus Agri-Marketing Inc. Web log, accessed January 2014, http://www.stratusresearch.com/blog07.htm.

Frisvold, G. B., T. M. Hurley, and P. D. Mitchell. 2009. "Adoption of Best Management Practices to Control Weed Resistance by Corn, Cotton, and Soybean Growers," *Agbioforum* 12:370-381.

Grube, A.H., D. Donaldson, T. Kiely, and L. Wu. February 2011. *Pesticides Industry Sales and Usage 2006 and 2007 Market Estimates,* U.S. Environmental Protection Agency.

Gustafson, D.I. 2008. "Sustainable Use of Glyphosate in North American Cropping Systems," *Pest Management Science* 64:409-416.

Harker, K.N., J.T. O'Donovan, R.E. Blackshaw, H.J. Beckie, C. Mallory-Smith, and B.D. Maxwell. 2012. "Our View," *Weed Science* 60(2):143-144.

Heap, I. 2014. *The International Survey of Herbicide Resistant Weeds,* accessed January 2014, http://www.weedscience.org/summary/home.aspx.

Heckman, J. 1979. "Sample Selection Bias as a Specification Error," *Econometrica* 47(1):153-161.

Hedrick, P.W. 2000. *Genetics of Populations,* second edition, Jones and Bartlett Publishers: Sudbury, MA.

Horowitz, J., R. Ebel, and K. Ueda. November 2010. *No-Till Farming is a Growing Practice,* EIB-70. USDA ERS.

Johnson, W.G., S.G. Hallett, T.R. Legleiter, F. Whitford, S.C. Weller, B.P. Bordelon, and B.R Lerner. November 2012. *2,4-D- and Dicamba-tolerant Crops – Some Facts to Consider.* Purdue University, ID-453-W.

Kaspar T.C., J.K. Radke, J.M. Laflen. 2001. "Small grain cover crops and wheel traffic effects on infiltration, runoff, and erosion." *Journal of Soil and Water Conservation* 56(2):160–164.

Lence, S.H. 2000. "Using Consumption and Asset Return Data to Estimate Farmers' Time Preferences and Risk Attitudes," *American Journal of Agricultural Economics* 82:934-947.

Livingston, M.J. April 2013. "Chapter 116: Economics of Pest Control," *Encyclopedia of Energy, Natural Resource and Environmental Economics,* J. Shogren, editor. Elsevier, Oxford.

Malik, J., G. Barry, and G. Kishore. 1989. "The Herbicide Glyphosate," *Biofactors* 2(1):17-25.

Mallory-Smith, C.A., and E. J. Retzinger. 2003. "Revised Classification of Herbicides by Site of Action for Weed Resistance Management Strategies," *Weed Technology* 17:605-619.

Miranda, M.J., and P.L. Fackler. 2002. *Applied Computational Economics and Finance*. MIT Press: Cambridge, MA.

Miranowski, J.A., and G.A. Carlson. 1986. "Economic Issues in Public and Private Approaches to Preserving Pest Susceptibility," *Pesticide Resistance: Strategies and Tactics for Management,* pp. 436–448. National Academy Press, Washington, DC.

Mueller, T.C., P.D. Mitchell, B.G. Young, and A.S. Culpepper. 2005. "Proactive versus reactive management of glyphosate-resistant or -tolerant weeds." *Weed Technology* 19:924-933.

National Research Council (NRC). 2010. *The Impact of Genetically Engineered Crops on Farm Sustainability in the United States*. National Academies Press, Washington, DC.

Norsworthy, J.K., S. Ward, D. Shaw, R. Llewellyn, R. Nichols, T. Webster, K. Bradley, G. Frisvold, S. Powles, N. Burgos, W. Witt, and M. Barrett. 2012. "Reducing the Risks of Herbicide Resistance: Best Management Practices and Recommendations," *Weed Science* 60 (Special Issue):31-62.

Osteen, C., and J. Fernandez-Cornejo. September 2013. "Economic and Policy Issues of U.S. Agricultural Pesticide Use Trends." *Pest Management Science* 69 (9):1001-1025.

Ostrom, E. 1999. "Coping with Tragedies of the Commons," *Annual Review of Political Science* 2: 493-535.

Ostrom, E., J. Burger, C.B. Field, R.B. Norgaard, and D. Policansky. 1999. "Revisiting the Commons: Local Lessons, Global Challenges," *Science* 284:278-282.

Price, A.J., K.S. Balkcom, S.A. Culpepper, J.A. Kelton, R.L. Nichols, and H. Schomberg. July/ August 2011. "Glyphosate-resistant Palmer amaranth: A threat to conservation tillage." *Journal of Soil and Water Conservation,* 66(4): 265-275.PurdueCorn and Soybean Herbicide Chart. https:// ag.purdue.edu/btny/weedscience/Documents/Herbicide_MOA_CornSoy_12_2012%5B1%5D.pdf

Reeves D.W. 1994. "Cover crops and rotations in Crops Residue Management," *Advances in Soil Science*. J.L. Hatfield, B.A. Stewart, editors, pp. 125-172. Lewis: Boca Raton, FL.

Reeves, D.W. 1997. "The role of soil organic matter in maintaining soil quality in continuous cropping systems," *Soil & Tillage Research* 43:131-167.

Rosenbaum, P.R., and D.B. Rubin. 1983. "The Central Role of the Propensity Score in Observational Studies for Causal Effects," *Biometrika* 70(April):41-55.

Ross, M.A., and C. A. Lembi. 2009. *Applied Weed Science Including the Ecology and Management of Invasive Plants*. Pearson Prentice Hall, NJ.

Ross, M.A., and D.J. Childs. April 1996. Herbicide Mode-Of-Action Summary. Report WS-23-W, Cooperative Extension Service, Purdue University, West Lafayette, IN.

Scott, B. A., and M. J. VanGessel. 2006. Delaware soybean grower survey of glyphosate-resistant horseweed (Conyza canadensis). *Weed Technology* 21:270–274.

32

The Economics of Glyphosate Resistance Management in Corn and Soybean Production, ERR-184
Economic Research Service/USDA

Sekhon, J.S. 2011. "Multivariate and Propensity Score Matching Software with Automated Balance and Optimization: The Matching Package for R," *Journal of Statistical Software* 42(7):1-52.

Shaw, D.R., M.D.K. Owen, P.M. Dixon, S.C. Weller, B.G. Young, R.G. Wilson, and D.L. Jordan. 2011. "Benchmark Study on Glyphosate-Resistant Cropping Systems in the United States. Part 1: Introduction to 2006-2008," *Pest Management Science* 67:741-746.

Smith, K. 2012. "Zero Tolerance to Glyphosate-Resistant Pigweed – An Arkansas Approach," Arkansas Soybean Promotion Board, United Soybean Board, Chesterfield, MO.

Union of Concerned Scientists. 2013. *The Rise of Superweeds – And What to do About it,* Policy Brief. Cambridge, MA, accessed December 2013, http://www.ucsusa.org/assets/documents/food_and_agriculture/rise-of-superweeds.pdf.

USDA ERS. 2012. *Commodity Costs and Returns, accessed August 2012,* http://www.ers.usda.gov/data-products/commodity-costs-and-returns.aspx.

USDA, National Agricultural Statistics Service (NASS). 2014. *Quick Stats,* accessed March 2014, http://quickstats.nass.usda.gov/.

USDA, Animal and Plant Health Inspection Service (APHIS). September 22, 2014. "Dow AgroSciences LLC; Determination of Nonregulated Status of Herbicide Resistant Corn and Soybeans," *Federal Register* 79(183):56555-6.

U.S. Environmental Protection Agency. September 1993. R.E.D. Facts: Glyphosate, EPA-738-F-93-011.

Vencill, William K., R.L. Nichols, T.M. Webster, J.K. Soteres, C. Mallory-Smith, N.R. Burgos, W.G. Johnson, and M.R. McClelland. 2012. "Herbicide resistance: toward an understanding of resistance development and the impact of herbicide-resistant crops," *Weed Science* 60(sp1):2-30.

Webster, T. M., and L. M. Sosnoskie. 2010. "The loss of glyphosate efficacy: a changing weed spectrum in Georgia cotton," *Weed Science* 58:73-79.

Weirich, J.W., D.R. Shaw, M.D.K. Owen, P.M. Dixon, S.C. Weller, B.G. Young, R.G. Wilson, and D.L. Jordan. 2011. "Benchmark Study on Glyphosate-Resistant Cropping Systems in the United States. Part 5: Effects of Glyphosate-Based Weed Management Programs on Farm-Level Profitability," *Pest Management Science* 67:781-784.

Wilson, R.G., B.G. Young, J.L. Matthews, S.C. Weller, W.G. Johnson, D.L. Jordan, M.D.K. Owen, P.M. Dixon, and D.R. Shaw. 2011. "Benchmark Study on Glyphosate-Resistant Cropping Systems in the United States. Part 4: Weed Management Practices and Effects on Weed Populations and Soil Seedbanks," *Pest Management Science* 67:771-780.

Wooldridge, J.M. *2010. Econometric Analysis of Cross Section and Panel Data,* second edition, MIT Press: Cambridge, MA.

Yancy, C.H. 2012. "Zero-Tolerance Zone," *Mid-South Farmer,* 17:8, 1 p.

Glossary of Key Terms

Annualized present value (APV) of returns is the return that, if received each year, would equal the observed present value of returns received over a particular time horizon. See also Appendix 1.

Common pool resources (CPRs) are resources that are particularly costly to exclude people from using, and from which use by one person reduces the availability to others (Ostrom et al., 1999). Both of these characteristics (non-excludability and subtractability) create a potential social dilemma in which users seeking shortrun objectives lead to outcomes that are not desirable from anyone's longrun perspective. As discussed in this report, weed susceptibility to herbicides can be considered a common pool resource.

Dominant/recessive traits are explained thus: a gene conferred to progeny (offspring) can be expressed (i.e., exhibited in observable characteristics or phenotype) when only one copy of the gene is present on one of the cell's paired chromosomes (if the trait is dominant) or may require that both copies of the gene carry the trait in order for it to be expressed (if the trait is recessive).

Fitness cost describes the notion that a weed that selects for herbicide resistance will probably not also be naturally selected to produce the most seeds or to grow best under a range of conditions.

Genetic engineering (GE) is a technique used to alter genetic material (genes) of living cells in vitro. A gene is a segment of DNA that expresses a particular trait such as insect resistance or herbicide tolerance. These modern techniques of agricultural biotechnology often speed the process of traditional plant breeding to confer other desirable characteristics to the crop.

Glyphosate is a broad-spectrum systemic herbicide active ingredient used to kill a broad range of weeds that compete with commercial crops. Glyphosate, known by many trade names including Roundup, is a highly effective herbicide.

Glyphosate resistance (GR) (see HR).

Herbicides are substances that control weeds and other plants. Herbicides are sold as mixtures of active ingredients (the biologically active component of the herbicide) with inert materials used to improve safety and facilitate storage, handling, or application.

Herbicide resistance (HR) is the inherited ability of a plant to survive and reproduce following exposure to a dose of herbicide normally lethal to the wild type. A population of a weed species that is usually controlled with the herbicide survives and reproduces; successive generations are no longer controlled with that herbicide.

Herbicide tolerant (HT) crops are commercial crops developed, usually via genetic engineering, to be unaffected by direct contact with specific herbicides, the most widely used being glyphosate.

Mode of action (MOA) of an herbicide is the mechanism by which an herbicide affects a plant at the tissue or cellular level. Herbicides with the same MOA will have the same translocation (movement) pattern and produce similar injury symptoms to the plant (Ross and Childs, 1996).

Weed scientists classify glyphosate as an EPSP synthase inhibitor, the only material with that MOA (Mallory-Smith and Retzinger, 2003). The other herbicides considered in the analysis of resistance

The Economics of Glyphosate Resistance Management in Corn and Soybean Production, ERR-184
Economic Research Service/USDA

management are acetochlor and atrazine. Atrazine, used on corn, is a photosystem II inhibitor, while acetochlor, used on corn and soybeans, is a very long fatty acid synthesis inhibitor. Another herbicide, s-metolachlor, is used on corn and soybeans and has the same MOA as acetochlor, but is not considered in this report. Other herbicides, with different MOAs, can also be used in glyphosate resistance management.

Pesticides are products used to prevent or manage pests such as weeds (herbicides), insects (insecticides), and plant pathogens (e.g., fungicides).

Propensity Score. The propensity score procedure seeks to ensure that individuals' characteristics—in the treatment and the control groups—that might influence group assignment and the outcome variable are not statistically different between the two groups. Researchers have shown that comparing mean outcomes between treatment and control groups made up of individuals who otherwise have very similar characteristics can reduce sample-selection bias.

Residual herbicides are usually applied after the crop emerges, remain in the soil, and control weeds for a period of time after application. Some are pre-emergence herbicides, because they are applied before weeds emerge. Examples of commonly used residual herbicides on corn and soybeans are acetochlor and s-metolachlor.

Resistance management practice (RMP) includes using tillage, using at least one herbicide MOA different from glyphosate's, rotating pesticides to delay resistance, planting a cover crop in the fall, rotating crops, scouting for weeds, keeping written records of weeds observed, adjusting planting dates to avoid weeds, adjusting crop plant density, mowing field edges, cleaning equipment between fields, and cultivating fields for weed control.

Appendix 1—Bio-Economic Optimization Model

Biological Model

The biological model we use in this study was recently used to examine the potential effects of alternative herbicide choices on glyphosate resistance in ryegrass, horseweed, and Palmer amaranth (Gustafson, 2008).[27] For each year t, the model relates the seed density s_t (seeds/square meter), glyphosate-resistance level rt (resistance-gene frequency), and herbicide choice ht to the annual weed density wt (weeds/square meter) and next year's initial seed density and resistance level.

Glyphosate resistance and susceptibility are conferred by genes x and X, respectively. There are three genotypes: a susceptible homozygote (XX), a heterozygote (xX), and a resistant homozygote (xx).[28] The resistance gene x is completely dominant,[29] so the heterozygote and the resistant homozygote survive glyphosate applications at the same rate. The frequencies of each genotype are $G_t=[(1-r_t)^2, 2r_t(1-r_t), r_t^2]'$, respectively. (Gt is a three-by-one vector.) For each genotype, the seed densities are $S_t=s_t\,G_t$, and $S^i=s^i\,G^i$ are the immigrant seed densities, where $G^i=[(1-r^i)^2, 2r^i(1-r^i), r^{i2}]'$.[30]

Prior to planting, $A_t=\{ep/(1+exp\{1.2[ln(s_t+s^i)-ln(d_{50})]\})\}(S_t+S^i)$ are the numbers of weeds of each genotype. The term in brackets is the fraction of weeds that potentially emerge, accounting for seed and density-dependent plant emergence: e is the fraction of seeds that germinate; p is the fraction of seedlings that emerge; and d_{50} is the number of seeds per hectare at which 50 percent germinate.

Weeds that survive after the crop is planted are $B_t=f^r\,A_t$, where

$$f^r = \begin{cases} 0.03 & h_t \in \{residual + glyphosate, all\ herbicides, residual + alternative\} \\ 1.00 & h_t \in \{glyphosate, glyphosate + alternative, alternative\} \end{cases}$$

is the fraction of horseweeds that survive application of the residual herbicide. We do not simulate resistance to the residual herbicide, so each genotype is reduced by the same fraction when the residual herbicide is used.

Horseweeds surviving post-emergence herbicide applications are $C_t=f^o.*B\_t,\ where\ .*$ denotes element-by-element multiplication, and

$$f^o = \begin{cases} f^g & h_t \in \{residual + glyphosate, glyphosate\} \\ f^a f^g & h_t \in \{all\ herbicides, glyphosate + alternative\} \\ f^a & h_t \in \{residual + alternative, alternative\}; \end{cases}$$

---

[27]Gustafson (2008) does not describe the biological model, known as the Herbicide Resistance Modeling System (HERMES); however, the author provided us with an electronic copy of the model.

[28]A genotype is the genetic makeup of an organism as opposed to its physical appearance (http://www.thefreediction-ary.com/).

[29]Genes conferred to progeny offspring can be expressed (called the "phenotype") when combined with another gene (dominant) or only when paired with a copy of the same gene (recessive). A homozygote has identical pairs of genes (alleles) for any given heredity characteristics.

[30]A fixed number of seeds with a fixed resistance level are assumed to land on the field uniformly from an external source each year to simulate the impact of nearby, wild horseweed populations, which are not selected for glyphosate resistance each year.

The Economics of Glyphosate Resistance Management in Corn and Soybean Production, ERR-184
Economic Research Service/USDA

where fg=[0.35, 1.00, 1.00]/ are the fractions of horseweeds that survive a glyphosate application, and fa=0.35(1.01) is the fraction of all horseweed genotypes that survive the alternative, post-emergence herbicide.[31]

$R_t = \left(C_t(2) + 2C_t(3)\right) / \left(2\left(C_t(1) + C_t(2) + C_t(3)\right)\right)$ denotes the glyphosate-resistance level for the surviving weeds. $C_t(1)$, $C_t(2)$ and $C_t(3)$ are the susceptible homozygotes, heterozygotes, and resistant homozygotes, respectively. The resistance level for the seeds produced by these plants is based on the fraction of the population that self-fertilize, s^f, and the fraction that mate randomly, $(1 - s^f)$. The resistance level also depends on the herbicide choice, which not only kills weeds but also reduces the ability of surviving weeds to produce seeds.

The numbers of self-fertilized seeds for each genotype are $S_t^s(1) = s^f s^p f^s(1)\left[C_t(1) + 0.25 C_t(2) f^c(2)\right]$, $S_{t+1}^s(2) = s^f s^p f^s(2)\left[0.5 C_t(2) f^c(2)\right]$ and $S_{t+1}^s(3) = s^f s^p f^s(3)\left[0.25 C_t(2) f^c(2) + C_t(3) f^c(3)\right]$; and the numbers of randomly-mating seeds are $S_{t+1}^m(1) = \left(1 - s^f\right) s^p f^s(1)\left[C_t(1)(1 - R_t) + 0.5 C_t(2)(1 - R_t) f^c(2)\right]$, $S_{t+1}^m(2) = \left(1 - s^f\right) s^p f^s(2)\left[C_t(1) R_t + 0.5 C_t(2) f^c(2) + C_t(3)(1 - R_t) f^c(3)\right]$, and $S_{t+1}^m(3) = \left(1 - s^f\right) s^p f^s(3)$ $\left[0.5 C_t(2) R_t f^c(2) + C_t(3) R_t f^c(3)\right]$.[32]

The fraction of seeds that survive herbicide choice h_t is given by:

$$f^s = \begin{cases} f^{rs} f^{gs} & h_t = residual + glyphosate \\ f^{rs} f^{as} f^{gs} & h_t = all\,herbicides \\ f^{rs} f^{as} & h_t = residual + alternative \\ f^{gs} & h_t = glyphosate \\ f^{as} f^{gs} & h_t = glyphosate + alternative \\ f^{as} & h_t = alternative; \end{cases}$$

where $f^{rs} = 0.5$ is the fraction that survive the residual herbicide, f^{gs}=[0.12, 0.87, 0.87]/ are the fractions of each genotype that survive glyphosate and $f^{as}=0.12(1.01)$ is the fraction that survive the alternative herbicide. Fitness costs of resistance reduce the number of seeds produced by the heterozygote, $f^c(2)=0.8$, and the resistant homozygote, $f^c(3)=0.6$, relative to the susceptible homozygote.

Finally, the seed densities at the beginning of t+1 are $s_{t+1} = \left[\sum_{i=1}^{3}\left(S_{t+1}^s(i) + S_{t+1}^m(i)\right)\right]$, and

$r_{t+1} = \left(S_{t+1}^s(2) + S_{t+1}^m(2) + 2\left(S_{t+1}^s(3) + S_{t+1}^m(3)\right)\right) / \left(2\sum_{i=1}^{3}\left(S_{t+1}^s(i) + S_{t+1}^m(i)\right)\right)$ are the initial glyphosate-resistance levels.

Economic Model

Returns (profits) received at the end of year t from the production of crop i_t (c for corn and s for soybean) is $\pi\left(h_t; s_t, r_t, i_t\right) = p^{i_t} y_t^{i_t}\left(w_t\right) - F^{i_t} - v\left(h_t\right)$, where p^{it} is a fixed, real price per bushel (2010 US$), $y_t^{i_t}\left(w_t\right)$ is crop yield, F^{it} is a fixed production cost, and $v(ht)$ is the cost of the herbicide choice. A grower that does not manage resistance maximizes annual profit, ignoring impacts on future seed densities and resistance levels. The grower that manages resistance maximizes the

---

[31]The relative fitness parameters for the susceptible homozygotes are based on the mean application rates for glyphosate from the 2010 ARMS data and LD50 values. The alternative, post-emergence herbicide is assumed to be 1-percent less effective than glyphosate in killing horseweeds. This assumption is based on econometric estimates of a model of damage abatement, originally developed by Chambers and Lichtenberg (1994), estimated using 2010 ARMS data.

[32]See Hedrick (2000) for a derivation of these equations.

37

The Economics of Glyphosate Resistance Management in Corn and Soybean Production, ERR-184
Economic Research Service/USDA

present value of profits received over an infinite horizon, $\sum_{t=0}^{\infty} \rho^t \pi\left(h; s_t, r_t, i_t\right)$, accounting for all impacts on future seed densities and resistance levels.

The annual weed density, $w_t = A_t + B_t + C_t$, is used to link the biological model to crop yield using Benchmark-Study data (Shaw et al., 2011) on A_t (weed density in spring prior to planting), B_t (weed density after planting and pre-emergence, herbicide use), C_t (weed density after the final post-emergence, herbicide application) plus weed density at harvest and corn and soybean yields.[33] (See box, "Data Sources.") Because crop growers use tillage and apply herbicides to control weeds and reduce yield losses, crop yield, annual weed density, herbicide use, and tillage are simultaneously determined. Indeed, Heckman's (1979) test rejects the hypothesis that weed density, herbicide use, and tillage intensity are exogenous for each cropping scenario (Appendix tables 1-1, 1-2).

To account for endogeneity and the separate effects of herbicide use and tillage on weed density, and then the effects of weed density on crop yield, two-equation systems are estimated for each cropping scenario using the generalized method of moments (GMM).[34] The GMM estimates indicate that corn and soybean yields declined with cumulative weed densities for each cropping scenario. Because the marginal impacts of weed density on corn and soybean yield are not statistically different across cropping scenarios, and because the overidentifying restrictions test rejects the exogeneity of the instruments in the continuous-soybean model at the 1-percent significance level, we base the marginal impacts of weed density for each model using the results for the crop-rotation scenario (Appendix tables 1-1, 1-2, 1-3).[35] For the corn-soybean scenario, soybean and corn yields are $y_t^s\left(w_t\right) = exp\left(4.0677 - 0.0028 w_t\right)$ and $y_t^c\left(w_t\right) = exp\left(5.3110 - 0.0047 w_t\right)$, respectively.[36] For the continuous cropping scenarios, soybean and corn yield are $y_t^s\left(w_t\right) = exp\left(3.9300 - 0.0028 w_t\right)$ and $y_t^c\left(w_t\right) = exp\left(5.2501 - 0.0047 w_t\right)$, respectively.

Finding Solutions

We solve for the managing resistance choices using dynamic programming, because it is the preferred method for solving dynamic optimization problems when information about the underlying state variables can be updated regularly. Because an analytical solution does not exist, it is necessary to use numerical methods. First, we specify a finite number of values each of the state variables can assume. The larger the number of values allowed, the more accurate the solution, but the longer it

---

[33]Data from the Benchmark Study (Shaw et al., 2011) are also used to check whether simulated and empirical weed and seed densities are consistent for each cropping scenario. The simulation results indicate this is the case and, in addition, that the simulated densities are consistent with other studies that have reported weed and seed densities under different scenarios (e.g. Davis and Johnson, 2008).

[34]The data are pooled because there are only 4 years of observations, and because each field was split into two halves, which effectively doubles the number of observational units.

[35]The GMM estimates are much more precise than the 2SLS estimates, and unlike the 2SLS estimates, the GMM estimates indicate a negative impact of cumulative weed densities on corn and soybean yields for each cropping scenario. Overidentifying restrictions tests do not reject exogeneity of the instruments for the GMM and 2SLS corn and soybean yield models for the corn-soybean scenario. Overidentifying restrictions tests reject exogeneity of the instruments for the 2SLS continuous-corn model at the 1-percent level but do not reject the exogeneity of the instruments for the GMM continuous-corn yield model at the 10-percent level. Overidentifying restrictions tests reject exogeneity of the instruments for both the 2SLS and GMM continuous-soybean models at the 1-percent level. Because the GMM coefficient estimates provide better in-sample forecasts for continuous soybean yield than the 2SLS and OLS estimates, the GMM estimates are used for the continuous-soybean model.

[36]The constants in all of the yield equations incorporate any statistically significant year and State, fixed-effects, as well as the mean-squared-error of the estimate.

takes to solve. We allow 51 values for the glyphosate-resistance level, between 1.0e-8 and 1.0, and 51 values for the initial seed density, between 0.1 and 6300, using the so-called Chebychev nodes (Miranda and Fackler, 2002).[37]

The choices of ignoring resistance (not managing resistance) are found by going through each possible value of the state vector and finding the herbicide choice that maximizes annual profit received at harvest. The choices for the case of managing resistance are found using Bellman's (1957) equation, $v(s_t, r_t, i_t) = \max_{h \in [1...6]} \pi(h; s_t, r_t, i_t) + \rho v(b(h; s_t, r_t, i_t))$, where $v(s_t, r_t, i_t)$ maps the state vector into the maximum present value of profit received over an infinite horizon starting in year t. Solving the Bellman equation involves finding a numerical approximation of the value function using an iterative procedure.[38]

The procedure is initiated by finding the coefficients of a trivariate, Chebychev polynomial that can be used to approximate the profit function at each state vector. The Bellman equation is then solved using that approximant as the initial guess for the second term on the right-hand side. This provides, for each state vector, a new estimate for the value function.

The coefficients of the trivariate, Chebychev polynomial that can be used to approximate the new estimate of the value function at each state vector are then recomputed. The Bellman equation is solved again using the updated approximant as the second term on the right-hand side, and the procedure is repeated until the value function on the last iteration does not differ by more than $1.0e-6, at any value of the state vector, from the value function on the second-to-last iteration.[39]

---

[37]The lower bound for the resistance-gene level is based on the initial resistance-gene level used by Gustafson (2008). The initial seed density is based on the minimum and maximum seed densities from the Benchmark Study (Wilson et al., 2011).

[38]The function b denotes the biological model (described above) that maps the current herbicide choice, h, and the current values of the state variables into next period's values of the state variables.

[39]See Miranda and Fackler (2002) for a description of how to numerically solve dynamic-programming problems. We use Fackler's code, which is available free on his website (http://www4.ncsu.edu/~pfackler/), to obtain the coefficients for the trivariate, Chebychev polynomials and to evaluate those polynomials.

The Economics of Glyphosate Resistance Management in Corn and Soybean Production, ERR-184
Economic Research Service/USDA

Generalized Method of Moments (GMM) estimates for the yield models for the corn-soybean rotation scenario

Coefficient	Corn			Soybean		
	Estimate	Standard error	Mean	Estimate	Standard error	Mean
Weed density equation						
Dependent variable	Weed density	-	25.69	Weed density	-	25.28
Standard error	36.93	-	-	37.38	-	-
Intercept	57.20 ***	16.124	-	-25.61	15.579	-
Year=2006	-17.79 ***	6.471	-			
Irrigated field			0.10	15.64	9.454	0.13
Treated field			0.50	-32.21 ***	8.217	0.50
Tillage intensity	-25.17 ***	8.967	0.54	0.59	6.738	0.51
Application index	-3.41	2.827	3.75	18.92 ***	4.950	3.43
Yield equation						
Dependent variable	ln(corn yield)		176.18	ln(soybean yield)		52.81
Standard error	0.22		-	0.22	-	-
Intercept	5.43 ***	0.055	-	4.06 ***	0.028	-
Year=2006	-0.22 ***	0.053	0.27	-0.01	0.037	0.25
Year=2007	-0.07 *	0.039	0.26	-0.06	0.039	0.26
Year=2008	-0.05	0.048	0.26	-0.06 *	0.034	0.24
State=IL	-0.03	0.040	0.26	-	-	0.22
State=IN	-0.12 ***	0.044	0.20	-	-	0.19
State=NE	-0.18 ***	0.049	0.24	-	-	0.24
Weed density	-4.75E-3 ***	1.52E-3	-	-2.78E-3 ***	9.27E-4	-
Field latitude	-	-	41.03	-	-	40.38
Restrictions test	X_8=9.85	-	-	X_{10}=15.97	-	-
Yield model intercept	5.31	-	-	4.07	-	-

Notes: Data are from 140 and 136 corn and soybean growers, respectively, in IN, IL, IA, MS and NE for 2006-2009 (Shaw et al., 2011). GMM estimates are for the two-equation system, and 2SLS estimates are for one equation. *, **, and *** indicate statistical significance at 10-, 5-, and 1-percent levels, respectively. Mean yields (not the mean of the natural log of yields) are reported. The instrumental variables include an intercept; dummy variables for years 2006, 2007, and 2008; State dummy variables for IL, IN, and NE; irrigation and treated-field dummy variables; and the field's latitude. Under the null hypothesis that the instruments are uncorrelated with the error term, X_{df} is asymptotically distributed chi-squared with *df* degrees of freedom (Wooldridge, 2010, p. 226). The null hypothesis that tillage intensity, the herbicide application index, and the weed density are uncorrelated with the error term can be rejected at the 10- and 5-percent levels for the corn and soybean models, respectively. Weed density (plants per square meter) is the annual sum of four measurements taken in the spring prior to planting, after planting and pre-emergence herbicide applications, after the final post-emergence herbicide application, and at harvest. Tillage intensity is 0 for no-till, 1 for conservation till, and 2 for conventional till. - = not applicable.

Source: USDA, Economic Research Service econometric estimates.

40

The Economics of Glyphosate Resistance Management in Corn and Soybean Production, ERR-184
Economic Research Service/USDA

Generalized Method of Moments (GMM) estimates for the yield models for the continuous-corn and continuous-soybean scenarios

	Corn			Soybean		
Coefficient	Estimate	Standard error	Mean	Estimate	Standard error	Mean
Weed density equation						
Dependent variable	Weed density	-	51.53	Weed density	-	93.07
Standard error	120.59	-	-	105.97	-	-
Intercept	209.48 **	84.21	-	-17.35	31.16	-
Irrigated field	104.09 **	45.57	0.48	-	-	-
Treated field	68.50 *	37.38	0.50	-	-	-
Tillage intensity	-2.92	32.13	1.31	-31.00 ***	8.80	0.47
Application index	-98.48 **	42.76	2.53	27.59 ***	7.70	4.05
Yield equation						
Dependent variable	ln(corn yield)	-	168.08	ln(soybean yield)	-	40.10
Standard error	0.37	-	-	0.36	-	-
Intercept	5.18 ***	0.04	-	3.86 ***	0.05	0.26
Year=2006	-	-	0.25	-	-	0.27
Year=2007	-	-	0.25	-	-	0.25
Year=2008	-	-	0.26	-	-	0.24
State=IL	-	-	-	-	-	0.20
State=IN	-	-	-	-	-	0.20
State=MS	-	-	-	-	-	0.36
State=NC	-	-	-	-	-	-
State=NE	-	-	0.56	-	-	-
Weed density	-1.33E-03 *	7.73E-04	-	-2.24E-03 ***	4.84E-04	-
Irrigated field	-	-	0.48	-	-	0.07
Treated field	-	-	0.50	-	-	0.50
Field latitude	-	-	41.94	-	-	36.59
Restrictions test	X_9=13.79	-	-	X_{10}=66.81 ***	-	-
Yield model intercept	5.25	-	-	3.93	-	-

Notes: Data are from 162 and 220 corn and soybean growers, respectively, in IL, IN, IA, MS, NC, and NE for 2006-2009 (Shaw et al., 2011). GMM estimates for two-equation systems and single-equation, 2SLS estimates are reported. *, **, and *** indicate statistical significance at 10-, 5-, and 1-percent levels, respectively. Mean yields (not the mean of the natural log of yields) are reported. Instrumental variables for corn include an intercept, dummy variables for years 2006, 2007, and 2008; a State dummy variable for NE; irrigation and treated-field dummy variables; and the field's latitude. Instrumental variables for soybean include an intercept, dummy variables for years 2006, 2007, and 2008; State dummy variables for IL, IN, and MS; irrigation and treated-field dummy variables; and the field's latitude. Under the null hypothesis that the instruments are uncorrelated with the error term, X_{df} is asymptotically distributed chi-squared with *df* degrees of freedom (Wooldridge, 2010, p. 226). The null hypothesis that tillage intensity, the herbicide application index, and the weed density are uncorrelated with the error term can be rejected at the 1-percent level for the corn and soybean models, respectively. Weed density (plants per square meter) is the annual sum of four measurements taken in the spring prior to planting, after planting and pre-emergence herbicide applications, after the final post-emergence herbicide application, and at harvest. Tillage intensity is 0 for no-till, 1 for conservation till, and 2 for conventional till. - = not applicable.

Source: USDA, Economic Research Service econometric estimates.

Appendix table 1-3
Bio-economic model parameters, base values, and sources

Parameter	Value	Source
Lower bound for seed density, s_t (seeds/m^2)	0.10	Wilson et al. (2011)
Upper bound for seed density, s_t (seeds/m^2)	6300	Wilson et al. (2011)
Lower bound for resistance level, r_t	1.0E-8	Gustafson (2008)
Upper bound for resistance level, r_t	1.00	Maximum
Emergence factor for seed, e	0.80	Gustafson (2008)
Density at which 50% of seeds germinate, d_{50}	5.0E+6	Gustafson (2008)
Emergence factor for seedlings, p	0.10	Gustafson (2008)
Incoming seed (seeds/hectare), s^i	10,000	Gustafson (2008)
Resistance level incoming seed, r^i	1.0E-6	No data
Self-fertilization fraction, s^f	0.95	Gustafson (2008)
Seeds produced per plant, s^p	32,665	Davis and Johnson (2008)
Reduction in seed production relative to XX for xX, $f^c(2)$	0.80	Gustafson (2008)
Reduction in seed production relative to XX for xx, $f^c(3)$	0.60	Gustafson (2008)
Relative fitness XX vs glyphosate, $f^g(1)$	0.35	Gustafson (2008)
Relative fitness xX and XX vs glyphosate, $f^g(2)$ and $f^g(3)$	1.00	Gustafson (2008)
Relative fitness all genotypes vs residual, f^r	0.03	Gustafson (2008)
Relative fitness all genotypes vs alternative, f^a	0.35	ERS estimates based on 2010 ARMS
Impact on seed production for all genotypes due to residual, f^{rs}	0.50	Gustafson (2008)
Impact on XX seed production due to glyphosate, $f^{gs}(1)$	0.12	Gustafson (2008)
Impact on xX and xx seed production due to glyphosate, $f^{gs}(2)$ and $f^{gs}(3)$	0.87	Gustafson (2008)
Impact on seed production for all genotypes due to alternative herbicide, f^{as}	0.12	ERS estimates based on 2010 ARMS
Annual discount factor, ρ	0.95	Lence (2000)
Application rate glypohsate (lbs ai/acre)	0.94	2010 ARMS
Application rate atrazine (lbs ai/acre)	0.98	2010 ARMS
Application rate acetochlor (lbs ai/acre)	1.18	2010 ARMS
Application rate residual (lbs ai/acre)	1.41	2010 ARMS
Price per lb ai glyphosate	6.49	2010 ARMS
Price per lb ai atrazine	6.70	2010 ARMS
Price per lb ai acetochlor	10.05	2010 ARMS
2010 marketing year corn price ($/bushel), p^1	5.18	USDA (2014)
2010 marketing year soybean price ($/bushel), p^0	11.3	USDA (2014)
Fixed production costs for corn, f^1	542.9	USDA (2012)
Fixed production costs for soybeans, f^0	379.79	USDA (2012)

Notes: XX = susceptible homozygote. xX = heterozygote. xx = resistant homozygote. ai = active ingredient.

Source: Various (see third column).

Appendix 2—Propensity-Score Matching Procedure

We use a multivariate, iterative matching algorithm that converges to the optimal matched sample by minimizing the largest discrepancies between the characteristics of individuals in the groups that are being compared (Sekhon, 2011). In the literature, the groups are typically referred to as the treatment group and the control group. In the example above, corn growers with GR weeds are in the treatment group, and corn growers without GR weeds are in the control group. We use one-to-one matching with replacement. If an individual in the treatment group matches more than one individual in the control group, the matched dataset includes the multiple matched control individuals, weighted to reflect the multiple matches.

We apply the method, using 2010 and 2012 Phase-II and Phase-III ARMS data, to estimate differences in several outcome variables of interest for six samples:

1. Corn growers who observed and did not observe a GR-weed infestation as of 2010,

2. Soybean growers who observed and did not observe a decline in the efficacy of glyphosate as of 2012,

3. Corn growers who used glyphosate by itself in 2010 and corn growers who used glyphosate with at least one different herbicide,

4. Soybean growers who used glyphosate by itself in 2012 and soybean growers who used glyphosate with at least one different herbicide,

5. Soybean growers who answered "yes" to the common pool resource (CPR) beliefs question and soybean growers who chose another answer, and

6. Soybean growers who answered "don't know" to the CPR beliefs question and soybean growers who chose another answer.

The propensity score is the estimated likelihood of belonging to the treatment group. For each sample, the propensity-score model includes an intercept, State dummy variables, the operator's age, whether the operator graduated from college, total acres planted to the surveyed crop on the farm, the fraction of total operated acres owned by the operation, the number of years since any GT crop was first planted on the surveyed field, and the number of times soybeans were planted on the surveyed field during the previous 4 years. A dummy variable indicating whether corn or soybean growers had observed GR weeds as of 2010 and 2012, respectively, is included in the last four samples. The propensity score is estimated using a standard, probit model for each sample.

The variables included (Appendix table 2-1) are (1) the propensity score, (2) the variables used to estimate the propensity score, (3) dummy variables indicating whether a decline in the efficacy of glyphosate had been observed, (4) the various seed types that were planted, (5) total acres planted to the surveyed crop on the farm (small, medium, large, very large), (6) whether the operation specialized in crop production, (7) the operator's occupation (grower, unemployed, off-farm worker, retired), (8) whether the operator was married, and (8) the operation's legal status (family operation, legal partnership, c-corporation or s-corporation).[40] Matching statistics for each sample are reported in Appendix table 2-1.

---

[40]The Florence (1933) first quartile, median, and third quartile were used to identify small, medium, large, and very large farms.

43

The Economics of Glyphosate Resistance Management in Corn and Soybean Production, ERR-184
Economic Research Service/USDA

Appendix table 2-1
Means for variables used in the propensity-score matching procedures

| Variable | Observed glyphosate-resistant weeds | | | | Used glyphosate by itself | | | | Answer to CRP beliefs question | | | |
| | Corn | | Soybean | | Corn | | Soybean | | Soybean | | Soybean | |
	Yes	No	Yes	No	Yes	No	Yes	No	Yes	All else	Uncertain	All else
Propensity score	0.149	0.142	0.506	0.502	0.427	0.422	0.507	0.501	0.540	0.535	0.287	0.285
AR	-	-	0.087	0.085	-	-	0.044	0.044	0.041	0.041	0.093	0.095
GA	0.126	0.126	-	-	0.029	0.026	-	-	-	-	-	-
IL	0.053	0.053	0.071	0.071	0.061	0.058	0.064	0.073*	0.093	0.083*	0.070	0.073
IN	0.095	0.095	0.072	0.073	0.029	0.029	0.076	0.062*	0.077	0.080	0.080	0.080
IA	0.074	0.084	0.100	0.099	0.087	0.102	0.073	0.072	0.083	0.078	0.095	0.095
KS	0.053	0.053	0.062	0.061	0.041	0.032	0.048	0.046	0.039	0.039	0.068	0.065
KY	0.063	0.063	0.029	0.029	0.044	0.055	0.039	0.039	0.025	0.041**	0.018	0.018
LA	-	-	0.020	0.020	-	-	0.009	0.009	0.035	0.034	0.038	0.038
MI	0.032	0.011	0.016	0.016	0.076	0.058	0.081	0.080	0.042	0.041	0.033	0.033
MN	0.053	0.053	0.056	0.056	0.128	0.146	0.060	0.050**	0.073	0.077	0.030	0.030
MS	-	-	0.048	0.048	-	-	0.031	0.024**	0.039	0.039	0.040	0.040
MO	0.074	0.074	0.088	0.094	0.035	0.047	0.054	0.054	0.050	0.049	0.063	0.068
NE	0.105	0.095	0.067	0.070	0.076	0.067	0.068	0.062**	0.075	0.092**	0.035	0.035
NC	0.116	0.105	0.040	0.038	0.029	0.026	0.016	0.013	0.035	0.030**	0.018	0.018
ND	0.032	0.053	0.042	0.042	0.096	0.087	0.096	0.096	0.062	0.062	0.043	0.033
OH	0.042	0.042	0.058	0.057	0.029	0.026	0.057	0.056	0.052	0.049	0.058	0.058
PA	0.032	0.021	-	-	0.015	0.015	-	-	-	-	-	-
SD	-	-	0.038	0.038	-	-	0.062	0.081**	0.050	0.050	0.100	0.100
TN	-	-	0.058	0.063	-	-	0.023	0.023	0.041	0.039	0.038	0.038
TX	0.042	0.042	-	-	0.111	0.093	-	-	-	-	-	--
VA	-	-	0.037	0.032			0.037	0.036	0.046	0.037	0.018	0.023
WI	0.000	0.000	0.010	0.010	0.012	0.009	0.064	0.080*	0.041	0.040	0.065	0.063
Operator's age	55.4	56.7	55.5	56.1	56.8	56.4	56.8	56.0**	55.0	55.7**	56.4	56.9
Operator graduated college	0.326	0.305	0.234	0.220	0.195	0.175	0.197	0.174**	0.260	0.215**	0.236	0.231
Share of total acres owned	0.418	0.451	0.404	0.391	0.583	0.595	0.509	0.469	0.406	0.380	0.460	0.414
Years GT crop use on farm	8.337	8.642	11.310	11.123	6.784	7.053	9.341	9.865**	10.890	11.097	10.995	11.048
Years soybean planted recently	1.537	1.568*	1.972	1.970**	1.041	1.067	1.697	1.738*	1.839	1.849**	1.925	1.895
Seed type												
GT-only crop planted	0.200	0.221	0.973	0.981	0.382	0.373	0.988	0.989	0.980	0.981	0.962	0.970*

Continued—

The Economics of Glyphosate Resistance Management in Corn and Soybean Production, ERR-184
Economic Research Service/USDA

Means for variables used in the propensity-score matching procedures—continued

Variable	Observed glyphosate-resistant weeds				Used glyphosate by itself				Answer to CRP beliefs question			
	Corn		Soybean		Corn		Soybean		Soybean		Soybean	
	Yes	No	Yes	No	Yes	No	Yes	No	Yes	All else	Uncertain	All else
Bt-only crop planted	0.189	0.189	-	-	0.122	0.122	-	-	-	-	-	-
GT and Bt crop planted	0.516	0.495	-	-	0.478	0.487	-	-	-	-	-	-
Other crop planted	0.095	0.074	0.025	0.019	0.017	0.017	0.011	0.011	0.019	0.019	0.038	0.030*
Non-Bt corn refuge size	0.098	0.112*	-	-	0.112	0.108	-	-	-	-	-	-
Enterprise size (acres)	292.0	268.3*	824.5	795.9	319.7	329.9	653.9	641.9*	766.2	735.5	725.5	742.3
Small (size < Q1)	0.768	0.800	0.625	0.641	0.799	0.816	0.720	0.726*	0.669	0.674	0.682	0.694
Medium (Q1 ≤ size < Q2)	0.168	0.158	0.209	0.186	0.122	0.108	0.145	0.146	0.187	0.173**	0.155	0.153
Large (Q2 ≤ size < Q3)	0.053	0.042	0.118	0.124	0.055	0.052	0.096	0.096	0.095	0.105	0.120	0.113
Very large (Q3 <= size)	0.011	0.000	0.049	0.049	0.023	0.023	0.040	0.032**	0.050	0.049	0.043	0.040
Crop farm	0.789	0.768	0.880	0.900	0.662	0.662	0.859	0.865**	0.875	0.889**	0.870	0.885*
Operator's occupation												
Farmer	0.853	0.863	0.886	0.894	0.837	0.848	0.866	0.871**	0.887	0.893*	0.862	0.872
Unemployed	0.021	0.021	0.008	0.003	0.020	0.020	0.008	0.007	0.005	0.004	0.013	0.013
Retired	0.042	0.042	0.052	0.048*	0.047	0.044	0.076	0.064*	0.046	0.031**	0.053	0.040
Off-farm worker	-	-	0.083	0.077*	-	-	0.094	0.098	0.090	0.091	0.095	0.093
Respondent was married	0.863	0.905	0.891	0.895	0.869	0.892	0.837	0.873**	0.861	0.878**	0.845	0.880*
Legal status of farm												
Family operation	0.863	0.895	0.709	0.718	0.834	0.837	0.761	0.790*	0.721	0.734**	0.732	0.757*
Legal partnership	0.074	0.063	0.163	0.168	0.090	0.090	0.145	0.120*	0.162	0.164	0.145	0.135
C-corporation	0.021	0.011	0.064	0.062	0.032	0.032	0.037	0.054**	0.054	0.052	0.070	0.050
S-corporation	0.042	0.032	0.054	0.049	0.035	0.035	0.042	0.028**	0.054	0.049	0.045	0.045
Observed GR weeds	-	-	-	-	0.041	0.023	0.341	0.340	0.494	0.453**	0.411	0.429

- = not applicable.

Source: USDA, Economic Research Service using data from the 2010 and 2012 Phase II and Phase III Agricultural Resource Management Survey.